Urban Sustainability

The Urban Sustainability Book Series is a valuable resource for sustainability and urban-related education and research. It offers an inter-disciplinary platform covering all four areas of practice, policy, education, research, and their nexus. The publications in this series are related to critical areas of sustainability, urban studies, planning, and urban geography.

This book series aims to put together cutting-edge research findings linked to the overarching field of urban sustainability. The scope and nature of the topic are broad and interdisciplinary and bring together various associated disciplines from sustainable development, environmental sciences, urbanism, etc. With many advanced research findings in the field, there is a need to put together various discussions and contributions on specific sustainability fields, covering a good range of topics on sustainable development, sustainable urbanism, and urban sustainability. Despite the broad range of issues, we note the importance of practical and policy-oriented directions, extending the literature and directions and pathways towards achieving urban sustainability.

The series will appeal to urbanists, geographers, planners, engineers, architects, governmental authorities, policymakers, researchers of all levels, and to all of those interested in a wide-ranging overview of urban sustainability and its associated fields. The series includes monographs and edited volumes, covering a range of topics under the urban sustainability topic, which can also be used for teaching materials.

Ali Cheshmehzangi

Resourceful Urbanism

Designing Regenerative Cities through Adaptive Reuse and Circular Innovation

Springer

Ali Cheshmehzangi
School of Architecture, Design & Planning
University of Queensland
Brisbane, QLD, Australia

ISSN 2731-6483 ISSN 2731-6491 (electronic)
Urban Sustainability
ISBN 978-981-95-4137-9 ISBN 978-981-95-4138-6 (eBook)
https://doi.org/10.1007/978-981-95-4138-6

This Springer imprint is published by the registered company Springer Nature Singapore Pte Ltd.
The registered company address is: 152 Beach Road, #21-01/04 Gateway East, Singapore 189721, Singapore

Four pillars of "*Resource Urbanism*" covered in this book – an image developed with the use of AI tools to support this central narrative – Brisbane, Sep 2025

Dedicated to people of the Global South who face racism, discrimination, forced displacement, injustice, exclusivity oppression, inequity, and inequalities of many forms.

Preface

> Yesterday I was clever, so I wanted to change the world.
> Today I am wise, so I am changing myself.
> —Rumi (Mo'lana), The well-known 13th-century Persian poet from Iran

For more than four decades we have been talking—endlessly and tediously—about "sustainable cities" and "sustainable development". We have filled libraries with frameworks, targets, indices, and glossy reports, each promising that sustainability is the horizon we must march towards. And yet, while the rhetoric has multiplied, our cities have grown more fractured, our industries more corrupted, and our ecological debt more staggering. Sustainability has become a word so overused, so abused, that it now functions as cover for inertia. It is the vocabulary of conferences and press releases, not of transformation.

Let us be blunt: the project of sustainable urbanism has failed. It has failed because it framed the future as an exercise in damage control, in doing "less bad" while leaving intact the extractive, inequitable logics of growth. It has failed because it placed its faith in high-tech fixes and consultant-driven frameworks that reproduce dependency rather than cultivate ingenuity. And it has failed because it never dared to address the deep corruption of the industries that profit from perpetual demolition, perpetual construction, and perpetual displacement.

This book is not about sustaining the "unsustainable". It is about revolution. It is about resourceful urbanism, an insurgent paradigm that takes the scarcity of our era not as a lament but as a design brief. Resourceful urbanism begins with the recognition that cities are not broken machines waiting for new parts; they are living ecologies of memory, relation, and possibility. And in this moment of planetary emergency, what cities need is not more spectacle, not more "smartness", not more hollow (and often endless) promises, but radical resourcefulness.

The cult of the "smart city" has been especially corrosive. For almost 15 years, we have been told that sensors, algorithms, and dashboards will save us; that if only we quantify every breath of air and every step on the sidewalk, efficiency will deliver us into utopia. What we got instead were cities beholden to corporate platforms, citizens reduced to data points, and planners chasing dashboards rather than justice. The smart city is not smart; it is surveillance dressed as progress. It is a fetish for gadgets that distracts from the fundamental truth: that technology without governance, without equity, without ecological humility, is just another form of extraction.

Equally bankrupt is the obsession with "growth-centric urbanism". The dream of endless expansion, of gleaming mega-projects rising from cleared land, is not vision but vandalism. It erases history, squanders embodied energy, and displaces communities in the name of "renewal"—and worst of all, for the glory of "development". Growth urbanism imagines constraint as an obstacle to bulldoze through, rather than a reality to work with. It clings to a fantasy of abundance even as planetary limits bear down on us. It is the logic of denial, one that guarantees ecological collapse.

Resourceful urbanism rejects both of these dead ends. It says: look around you. Look at the vacant warehouses, the underutilized corridors, the informal economies, the skills hidden in plain sight, and many other arcane realities of our contemporary society. Look at the communities organizing, the waste pickers recycling, the residents fighting for land, the ecosystems regenerating in neglected edges. These are the true laboratories of the future city, not the corporate tech estates or the glossy masterplans.

The evidence is here. In London, residents reclaimed derelict riverside land at Coin Street and turned it into cooperatives and community spaces. In Fiji, families built cyclone-resilient homes through sweat equity in Koroipita. In Denmark, industries stitched together their waste flows into Kalundborg's symbiotic metabolism. In Pune, waste pickers built a cooperative that redefined the circular economy from below. These are not "best practices" to be cut-and-pasted; they are radical acts of ingenuity that prove constraint can be the engine of regeneration.

Resourceful urbanism is not nostalgia for pre-industrial simplicity. Nor is it techno-utopian delusion. It is a third path—messy, creative, autonomous, and regenerative. It redefines what we mean by resources: not just steel, concrete, kilowatts, and joules, but also cultural memory, social networks, underused land, and adaptive capacity. It insists that governance is as important as design, that circularity is cultural as much as technical, that multifunctionality is the art of making spaces do more than one thing at once.

This book builds on my recent conception of the "*Advanced City*". Search for the term and you will get to know more of this new conception, or what I'd like to refer to it as my "urban revolution". Whether loud or silent, practical or left in theories, this conception leads to a new beginning, a chance to think and rethink repeatedly until we get back to our right path(s) again. If the twentieth century gave us the "modern city" and the twenty-first threatened to trap us in the pseudo-logic of the "smart city", the Advanced City is the choice we must make. It is not advanced because it has more gadgets, but because it has evolved in consciousness. It is people-centric, and it has "evolved by choice". It is advanced because it treats limits not as restrictions but as the starting point for creativity. It is advanced because it integrates governance, materials, ecosystems, and everyday practices into a regenerative whole. The Advanced City is resourceful at its core.

We are out of time for incrementalism. The climate crisis is accelerating, inequality is deepening, and geopolitical fractures are widening. To keep chanting "sustainable cities" is to cling to a broken promise. To keep chasing "smart cities" is to outsource our futures to corporations. We need new paradigms, urgent, grounded, and radical. Resourceful urbanism is one such paradigm. It does not pretend that cities can escape constraints; it shows how they can thrive because of them.

This preface is written not to comfort but to provoke. I want readers to feel the urgency, the anger, and the possibility. We cannot wait for perfect plans or endless budgets. We cannot trust industries that profit from demolition or platforms that profit from data extraction. The responsibility is ours, as designers, planners, architects, policymakers, and citizens, to reclaim the city as a space of ingenuity, solidarity, and adaptive creativity. It is time to innovate rightly, justly, and purposefully.

The chapters that follow are not just case studies; they are signals. They are the sparks of a revolution already underway. From Medellín's Library Parks to Yokohama's multifunctional landscapes, from Sydney's Goods Line to Cape Town's silo-turned-museum, from Pune to Kalundborg, the future is being built in fragments of resourcefulness across the globe. The task now is to connect these fragments into a movement, to turn isolated brilliance into systemic transformation.

This is not a book about sustainability as damage control. It is a book about regeneration as destiny. It is not about smartness as gadgetry. It is about wisdom as practice. It is not about growth for its own sake. It is about flourishing within limits.

We have been talking for 40 years. Enough. It is time to act, to design, to govern differently. We are living through perhaps the darkest moment of leadership in modern history, surrounded by maniacal figures who parade as "civilized", yet sow only chaos, disruption, and destruction. Their negligence and inaction have pushed us into a disastrous 2025, a year scarred by more than 60 conflicts and wars erupting across the globe. They led us to destruction, decay, decline, and outright loss of faith in so-called humanity. But now we are at the crossroads. We have to choose to be on the right side, not on their side! It is our chance to reimagine what we could do better, rethink our doings, and reflect on our downfall. It is indeed time to build cities and communities that are advanced not because they consume more, but because they regenerate more. It is time to make resourceful urbanism the new common sense. Let's stop thinking and act a bit more!

Brisbane, Australia Ali Cheshmehzangi

Acknowledgements

I would like to acknowledge and thank all those who continue to inspire me, whether through thoughtful discussions, spontaneous one-to-one conversations, or the shared energy of events and gatherings we attend regularly. These encounters, both formal and informal, create spaces where ideas are sparked, challenged, and reshaped. They offer me the chance to absorb perspectives, to think and rethink, to react and then to act, often setting in motion the flow that carries me into long evenings and late nights of passionate writing. Every exchange, every place, and every person encountered becomes part of the fabric that shapes my work. To acknowledge them is to recognize that no book is created in isolation; each one is born out of a constellation of people, moments, and places that make the journey meaningful and the ideas enduring.

Competing Interests The author has no competing interests to declare that are relevant to the content of this manuscript.

About the Book

Resourceful Urbanism reimagines how cities can thrive through constraints, whether environmental, economic, or spatial, by harnessing ingenuity, local knowledge, and innovative design. This book focuses on how adaptive reuse, circular thinking, and participatory governance might lead to urban resilience rather than high-tech fixes or significant advancements. It makes the case that being "resourceful" in the urban context means changing our values for things like relationships, materials, ecosystems, and lived experience in addition to getting more done with less.

The transformative tactics that influence the urban futures of both Global North and Global South contexts are examined in this book in a current and fact-based manner. It combines several approaches, i.e., formal and informal, grassroots and institutional, to show how locally rooted activities may make cities more regenerative, inclusive, and adaptable. The book is structured around four central themes that collectively define resourceful urbanism: "Community-Led Innovation", "Adaptive Reuse and Retrofitting", "Circular Resource Management", and "Multifunctional Urban Spaces".

Each theme is illustrated through two in-depth case studies, one from a developed context and one from a developing context, showing how contextually tailored interventions can drive systemic transformation.

Contents

About the Author

Ali Cheshmehzangi is Professor and Head of the School of Architecture, Design and Planning (ADP) at the University of Queensland. He is the Brisbane Olympics Committee member for Games Legacy and Growth Sub-Committee as well as the expert member of the 2032 Games led by UQ. He has been in the World's top 2 percent field leaders since 2021, recognized by Stanford University. He is among the top 20 global scholars in the urban sustainability research area.

Ali has a wealth of experience, expertise, and a strong commitment to innovation and excellence in the fields of architecture, design, and planning, the three core disciplines of his School at UQ. With a career spanning over two decades, he has made significant contributions to the academic and professional communities, with a focus on sustainable and environmentally conscious design. Prior to joining UQ, Ali held several strategic leadership and senior managerial roles, such as Vice-President for International Engagement and Global Partnership, Founding Director and Head of the Center for Innovation in Education and Research, Head of the Department of Architecture and Built Environment, Founding Director of the Urban Innovation Lab, Director of a university-wide Teaching and Learning platform, Director of the International Research Network for Rural and Urban Development, Founder and Head of Research Group for Sustainable Built Environment, co-director of university-wide research priority areas, Director of Center for Sustainable Energy Technologies, and Interim Director of Digital Design Lab. So far, Ali has published over 500 journal papers, articles, conference papers, book chapters, and reports. He has more than 50 academic book, alongside his many poetry collections that he has written in Parsi/Persian and English. Some of his previous academic books have received awards at the national, provincial, and municipal levels. He has also received international awards and recognition for his research on urban resilience studies and sustainability research, as well as a 2018 Vice-Chancellor's award for his impactful contribution to higher education.

Chapter 1
Beyond Scarcity: Reimagining Urban Futures Through Resourceful Innovation

1.1 The Urban Paradox: Abundance Within Constraints

Cities are humanity's greatest invention and its most dangerous gamble. They are hubs of wealth, creativity, and opportunity, but they are also the epicenters of environmental breakdown, fragility, and inequality. This paradox is the reality of the twenty-first century and is not only rhetorical. Cities produce almost 70% of the world's carbon emissions and use more than three-quarters of the world's energy [1, 2]. They consume materials, water, and land at previously unheard-of speeds. However, cities are also the places where creativity, culture, and group effort are most strongly expressed, and where the answers to social disintegration and climate change must be developed.

The paradox is sharpened by the fact that cities embody both abundance and scarcity simultaneously [3]. On the one hand, they provide unparalleled access to networks, services, and employment opportunities. However, they are rife with profound inequality: squalid informal settlements border opulent high-rise communities; expansive freeways suffocate youngsters without safe routes to school; mountains of food waste rise as others go hungry. Some people perceive what seems to be abundance as scarcity, and these two realities frequently coexist side by side inside the same metropolitan fabric.

Looking outward, hoping that technical advancements or vast new infrastructure can save cities from their inconsistencies, has frequently been the prevailing reaction to this dichotomy. A prime example is the "smart city," which, although being marketed as a seamless digital utopia, all too frequently offers little more than monitoring systems, elite app-based services, and glitzy displays of innovation that fail to solve structural injustices [4]. Similarly, the reasoning behind "tabula rasa" rehabilitation, which involves demolishing existing structures in order to construct new ones, replicates the cycles of dispossession and carbon-intensive building. In

A. Cheshmehzangi, *Resourceful Urbanism*, Urban Sustainability,
https://doi.org/10.1007/978-981-95-4138-6_1

addition to failing to address the conundrum, these tactics make it worse by widening gaps and hastening ecological overshoot.

The foundation of resourceful urbanism is a new idea: limitations are catalysts to be used rather than barriers to overcome. Scarcity is the raw material for creativity, not its enemy. Under strain, some of the most resilient urban shapes have developed over time. Despite their precarities, informal communities exhibit remarkable spatial creativity and adaptive reuse. Human-scale designs that last for generations can be seen in historic quarters that were constructed with little resources. The most durable networks of collaboration, recycling, and shared stewardship are frequently formed by communities that are experiencing economic hardship. Thus, the city's dilemma also presents an opportunity: if viewed from a new angle, its limitations might spur regenerative creativity.

The misconception that sustainability is inevitably costly or technologically complex must be debunked in order to achieve this. In actuality, low-tech, gradual, and very social urban developments frequently have the biggest effects [5]. A cooperative that formally integrates waste-pickers into a municipal recycling system, a community garden that repurposes idle land to feed people, or the conversion of former industrial buildings into cultural centers—these are not band-aid solutions; rather, they are the cornerstones of regenerative cities. They show that the elements of change are already present in urban life, including unused areas, trash, social knowledge, and natural processes that continue to exist even in damaged environments.

In this sense, cities are more like dynamic workshops—messy, layered, and alive—where new tools are continuously produced under strain, rather than being laboratories in the sterile scientific sense. They are abundant in relationships, histories, and untapped capacities in addition to infrastructure and economic flows. When seen in this light, the city's paradox changes: its excessive emissions and consumption are the precise reason it must and may serve as a catalyst for change on a global scale.

A conceptual key to transforming this dilemma into opportunity is provided by resourceful urbanism. The focus on constraint as a design brief asks how to develop with and through scarcity rather than how to get around it. Informal networks are reframed as knowledge repositories rather than as governance failures [6]. It views abandoned infrastructure as canvases for adaptive reuse rather than as urban trash. It maintains that social inclusion and ecological restoration are interconnected tactics rather than opposing goals.

The urban paradox, then, is not a trap but a challenge. In addition to being the primary cause of global problems, cities are the only scale at which group remedies may spread quickly enough to have an impact. To embrace the messy reality of urbanism and reframe it as a hub for creative ingenuity, it is necessary to confront abundance within limitations [7]. This book makes the case that by doing this, we may shift from extractive to regenerative models of development—cities that not only endure within their boundaries but flourish because of them.

1.2 Defining Resourceful Urbanism: A Paradigm Shift

Resourceful urbanism is not a slogan for thrift, nor a plea for endless efficiency. It is an appeal to radically rethink what constitutes a resource in urban life. Steel, cement, energy, and water are examples of resources that are measured in tons, kilowatts, or cubic meters, according to the mainstream development narrative. According to this reasoning, urban resilience entails obtaining more resources and handling them with higher technical accuracy. However, this reductionist perspective ignores the most important qualities that contribute to cities' resilience, adaptability, and humanity. The concept of resources is broadened by resourceful urbanism to encompass natural systems, social capital, cultural memory, adaptive informal economies, and underutilized urban areas. This perspective transforms the city's metabolism into a circulation of connections, customs, and opportunities in addition to a flow of resources and energy.

Dispelling the myth that sustainable urbanization can be achieved solely through technology is the first step in this paradigm shift. The idea that data, sensors, and automated control systems may optimize urban life into seamless efficiency is the foundation of the techno-utopian vision of the "smart city". Although these instruments can be useful, they frequently strengthen centralized authority, widen disparities, and bind cities to costly infrastructure that is vulnerable to crises. The value of walkable neighbourhoods cannot be replaced by smart traffic systems, and a digital waste-tracking platform is worthless without a community that is prepared to sort, recycle, and reuse its garbage. The smart city paradigm confuses tools for aims by placing technology above society.

Equally insufficient is the nostalgia that longs for a return to pre-industrial simplicity. Romantizing the past runs the risk of ignoring the reality of population increase, complex economics, and global interconnectedness, even though local craft traditions, historic textiles, and low-energy building methods all teach valuable lessons. Cities cannot be reduced to villages, nor should they aim to undo the tangible advancements that have helped millions of people escape poverty. To create futures that are grounded, hybrid innovations rather than regressions or rogue techno-futures, it is necessary to weave the lessons learned from the past with the demands of the present.

A third route is suggested by resourceful urbanism, which employs creativity and adaptive reuse as its main tactics and welcomes complexity rather than avoiding it. This method recognizes that every city has latent capacities that are just waiting to be discovered. A drainage channel is more than just infrastructure; a neighbourhood cooperative is a governance innovation rather than a marginal one; and an abandoned factory is not a dead site but rather a potential cultural anchor. Instead of relying on outside solutions, cities may revitalize themselves by redefining these underutilized assets as resources.

This worldview is centered on circularity. Resourceful urbanism insists on cycles of recovery and reinvention in place of linear systems of extraction, consumption, and disposal. Buildings should be modified rather than demolished, materials should be recycled, waste should be converted into inputs, and energy systems should be redesigned as integrated metabolisms [8]. Circularity, however, is a social as well as a technical concept in this context. Additionally, concepts, methods, and abilities need to be shared. The ability of a youth collective to administer neighbourhood Wi-Fi, an elderly resident's familiarity with ancient flood management practices, or a group of migrants operating a cooperative network of small companies are all equally important components of the city's circular economy as solar panels or recycling facilities.

Fundamentally, resourceful urbanism opposes extractive connections between communities and ecosystems. Extractive models view labour as disposable, buildings as disposable, and nature as a raw material. Cultural erasure, socioeconomic inequity, and environmental collapse are the outcomes. Regenerative relationships, on the other hand, acknowledge interdependence: ancient marketplaces as sources of identity and livelihood, waste-pickers as environmental stewards, and wetlands as flood defenders. Such acknowledgment necessitates not just new laws but also a cultural change in values, moving away from growth as defined by consumption and toward prosperity as measured by ecological health, equity, and resilience [9].

Because of this, resourceful urbanism can be both a critique and a positive substitute. It challenges popular conceptions that link advancement to technology-driven masterplans, speculative buildings, or megaprojects [10]. It reveals how these models all too frequently speed up ecological degradation, neglect the reality of informal life, and alienate communities from decision-making. However, it also offers beneficial substitutes, such as circular systems that lessen reliance on extraction, participatory governance that encourages grassroots creativity [11], multipurpose areas that increase the value of scarce land, and adaptive reuse that respects embodied carbon.

In addition to its scope, resourceful urbanism stands out for its emphasis on possibility. It sees possibility where others see ruin. Relationships and memories provide abundance where politicians see shortage. Developers identify opportunities for renewal where they observe dereliction. It suggests that crises can inspire creativity rather than dismissing them. It is a paradigm shift because it redefines what is valuable and deserving of being built atop, rather than just modifying outdated frameworks.

This idea will be shown through actual case studies in the upcoming chapters, showing that resourceful urbanism is a lived activity rather than an abstract theory. The evidence is clear: resourceful urbanism is becoming a new way of doing things in a variety of contexts. Examples include community-led infrastructure in the Pacific, adaptive reuse of industrial silos in South Africa, circular industrial symbiosis in Denmark, and multifunctional ecological landscapes in Japan. It's not about "doing more with less" in the austere sense; rather, it is about realizing that cities already have more than enough resources to rebuild themselves, if we can change the way we think and act.

This paradigm change is a practical necessity rather than an ideal. Creative, inclusive, and regenerative alternatives are needed to address the crises of social fragmentation, resource shortage, and climate change. That is exactly what resourceful urbanism provides: a framework that reinterprets resources, welcomes complexity, and maintains that the most potent solutions are those that emerge naturally from the limitations. With creativity as our most plentiful renewable resource, it challenges us to rebuild the city from the ground up rather than to abandon it.

1.3 The Imperative for Regenerative Cities

The next step is to understand why regenerative cities are now necessary rather than optional if resourceful urbanism transforms scarcity into opportunity. The scope and gravity of our current issues necessitate a fundamental shift in the way cities are planned, run, and inhabited. Rising oceans, more intense heat waves, and record-breaking floods are all signs that climate change is becoming a reality rather than a remote threat. Social inequality is a systematic fault line that weakens unity and undermines governance, not a regrettable byproduct of growth [12]. Depletion of resources is not a theoretical prediction; it is felt on a daily basis in the form of global supply shocks, declining water levels, and collapsing soils. Cities are at the center of these conflicting forces, acting as both perpetrators and victims.

Therefore, it is essential to make cities regenerative rather than just more efficient or less harmful. Beyond minimizing damage, regeneration entails actively mending natural systems, fostering communal ties, and facilitating cycles of renewal. In addition to lowering emissions, a regenerative city replenishes wetlands that store carbon and provide flood protection. Instead of separating populations, it builds dwellings in ways that recycle pre-existing materials and foster social trust. Rather than entangling communities in exploitative dependence, it uses distributed systems to provide energy.

This problem cannot be solved by disciplinary silos. Regeneration requires the cooperation of planners, architects, engineers, ecologists, sociologists, and economists. Community-led innovation is no longer an afterthought but the foundation of resilience; circular resource flows are no longer specialized experiments but rather essential to urban economies; and adaptive reuse of buildings is no longer an architectural preference but a climate requirement. Integration is necessary for regeneration, and integration calls for creativity.

The multipurpose landscapes of Yokohama, the industrial symbiosis of Kalundborg, the trash cooperatives of Pune, and the Library Parks of Medellín all show that regeneration is a real practice rather than a fantasy. They demonstrate how natural landscapes can also serve as civic commons, how waste-pickers can become environmental managers, how industrial byproducts can be turned into mutually beneficial uses, and how cultural infrastructure can also serve as a means of reducing violence.

The imperative is clear: cities need to treat restrictions as design briefs rather than as justifications for stagnation. The goal of regenerative urbanism is to thrive within the bounds of the earth, not to survive on the periphery. It starts with acknowledging that the future of cities is the future of the world, and it is both an ethical duty and a workable strategy.

1.3.1 Disciplinary Convergence and Contemporary Relevance

In addition to its tangible accomplishments, resourceful urbanism is significant because it challenges and reshapes the disciplinary boundaries that have traditionally limited urban activity. Planners create master plans, architects build structures, engineers manage infrastructure, economists advise on growth, and environmental scientists predict dangers. These days, modern cities are usually divided into different areas. Although each field offers expertise, the division of knowledge has all too frequently resulted in disjointed outcomes, such as housing developments that are not connected to transportation, flood defences that are constructed without taking ecosystems into consideration, and economic zones that are developed without any social or cultural underpinnings. A patchwork of initiatives that fail to capture the interwoven reality of urban life is the end consequence.

This compartmentalized reasoning is contested by resourceful urbanism. It maintains that the problems cities face—inequality, resource depletion, social fragmentation, and climatic breakdown—are systemic and hence call for systemic solutions. These challenges cannot be handled by one profession alone. Rather, resourceful urbanism builds ties between disciplines, between formal and informal systems, and between grassroots activism and institutional frameworks [13]. Its strength is not in favouring one type of expertise over another, but rather in combining them into methods that are context-specific and adaptive.

Take adaptive reuse, for instance. In addition to architectural creativity, planning permissions, economic models for operation, ecological studies for material reuse, and most importantly, community involvement to determine the building's intended uses are all necessary to transform a vacant warehouse into a cultural center. Such a project is co-produced through convergence rather than being owned by a single discipline in resourceful urbanism. Similarly, technical infrastructure, social organization, governance structures, and cultural acceptance are necessary for cyclical resource management [14], whether it is community-based recycling in Pune or industrial symbiosis in Kalundborg. It is a living system that combines engineering, economics, sociology, and politics.

This convergence involves linking the formal and informal spheres in addition to professional disciplines. Informal economies are not peripheral to urban life in a large portion of the Global South. Community house builders, waste pickers, and unofficial transit providers offer services that states are unable or unwilling to provide. Conventional planning frequently views these behaviours as issues that need to be eliminated. Rather, resourceful urbanism acknowledges them as sources of creativity that need to be included into official government structures [15, 16]. Instead of romanticizing informality, the outcome is a practical blending: traditional ecological knowledge is incorporated into formal environmental policy, garbage cooperatives are acknowledged in municipal contracts, and informal settlements are improved through participatory planning.

Resourceful urbanism also speaks to relevance in the modern day. Whether we like to accept it or not, we live in a time of convergence: social unrest breaks out where inequality and spatial segregation meet, pandemics highlight the connections between housing, health, and mobility, and climate shocks disregard professional borders. The framework and vocabulary offered by resourceful urbanism enable proactive rather than reactive responses to this interconnection. Instead of stumbling through complexity, it enables cities to anticipate it.

In practice, this entails developing cross-domain translators, institutions open to trying hybrid governance, and communities with the capacity to co-produce solutions. This book's examples, which range from Yokohama's multipurpose landscapes to Medellín's Library Parks, show how creative urbanization flourishes where disciplines meet. Cultural workers teamed up with ecologists, municipal engineers with farmers, and architects with social entrepreneurs. New types of institutional practice and inventive infrastructure were the results.

In conclusion, disciplinary convergence is embodied in resourceful urbanism as both a requirement and an opportunity. It demonstrates that resilience is found in the ability to weave across deeper silos of expertise, to adaptably blend the formal and informal, and to match technological design with social imagination. There is no denying its relevance today: in a century characterized by systemic crises, creative urbanism provides the connective tissue needed to make cities not only liveable but also regenerative.

1.3.2 Aims and Objectives: Rewriting Urban Futures

This book starts with the belief that creativity based on limitations, rather than boundless resources or innovative technical solutions, will be the key to rebuilding twenty-first-century cities. With conviction and clarity, Resourceful Urbanism aims to demonstrate how scarcity itself becomes a design brief—how professionals, organizations, and communities are already turning what seems like a deficit into opportunity. In addition to documenting these activities, the main objective is to present them as a cohesive paradigm that has the power to reshape urban futures.

Documenting and analyzing transformation practices under constraint is the primary goal. This book explores the transformation of abandoned rail lines, abandoned grain silos, informal settlements, or waste economies into resilient, inclusive, and culturally vibrant infrastructures through eight case studies. As seen by projects that transform defunct infrastructure into public areas, insecure labour into cooperative service systems, or delicate ecologies into multipurpose landscapes, resourceful urbanism is not merely theoretical but incredibly tangible. Each illustration demonstrates how creativity is applied in situations with constrained resources, inadequate governance, or pressing environmental concerns. By juxtaposing different methods, the book highlights common techniques, recurring problems, and emergent tactics, going beyond isolated success stories.

The second objective is to bridge Global North and Global South experiences into a shared framework of ingenuity. All too frequently, urban discourse treats "developing" and "developed" cities as having different levels of technological advancement and inadequacy. This dichotomy is rejected by resourceful urbanism. Rather, it demonstrates how a community cooperative in Pune can teach about circularity just as much as a Danish industrial symbiosis, or how Sydney's Goods Line and Medellín's Library Parks can transform underutilized areas into public infrastructure. According to the comparative approach, information flows must be two-way: behaviours in the North can be reframed by what comes from the South, and vice versa. The book presents resourceful urbanism as an international dialogue where many settings equally contribute to the rethinking of urban practice.

The third goal is to provide communities, politicians, and practitioners with practical solutions. This book presents critical conceptual arguments, although it is not meant to be a strictly theoretical exercise. It serves as a guide for converting case study lessons into tactics that may be used in other contexts. It illustrates the tools that enable creative ideas to be implemented, such as incremental phasing, participatory governance structures, and community land trusts, for urban planners. It outlines for policymakers the enabling conditions—institutional collaborations, financial incentives, and legal frameworks—that make such models possible. For communities, it emphasizes self-organization, collaborative management, and cultural innovation methods that restore agency in constructing urban landscapes.

Importantly, this book also aims to address the broader discussion of urban futures. The capital-intensive, technologically driven ideals of smart city platforms, mega-infrastructure, and speculative redevelopment continue to dominate mainstream models. Although they might play a part, they run the risk of escalating inequality and ignoring the abundance of creativity already found in local communities. To view cities as living ecologies where adaptive reuse, circular flows, and multipurpose spaces already exist as seeds of regeneration, resourceful urbanism advocates for a fundamental change in perspective.

Lastly, this book aims to establish a comparative, global framework that bridges the divide between the North and South, document and theorize practices that transform scarcity into opportunity, and provide tools that practitioners, policymakers, and citizens can use in the pressing task of creating regenerative cities. It has an unashamed critical and unapologetic practical purpose. By the end, readers will have a better understanding of resourceful urbanism and why it is essential for rewriting the course of urban futures in the face of social fragmentation and ecological constraints.

1.4 Navigating This Book: A Journey Through Urban Innovation

This book does more than just list projects and practice technical arguments. It provides an exploration of what it means to create cities that flourish in the face of constraint, encompassing a philosophical, practical, and ultimately political journey. Resourceful Urbanism is not portrayed as a collection of impersonal rules that should be implemented arbitrarily or as a plan. It is structured instead on four thematic pillars that serve as radical lenses through which to view the city (see Fig. 1.1):

- **Community-Led Innovation,**
- **Adaptive Reuse and Retrofitting,**
- **Circular Resource Management, and**
- **Multifunctional Urban Spaces.**

Together, these four themes carve out a new conceptual terrain, positioning resourcefulness not as a peripheral tactic but as the heart of regenerative urbanism.

These pillars are described in the following sections as provocations rather than as fixed groupings. Both are aspirational and analytical. Each challenges the

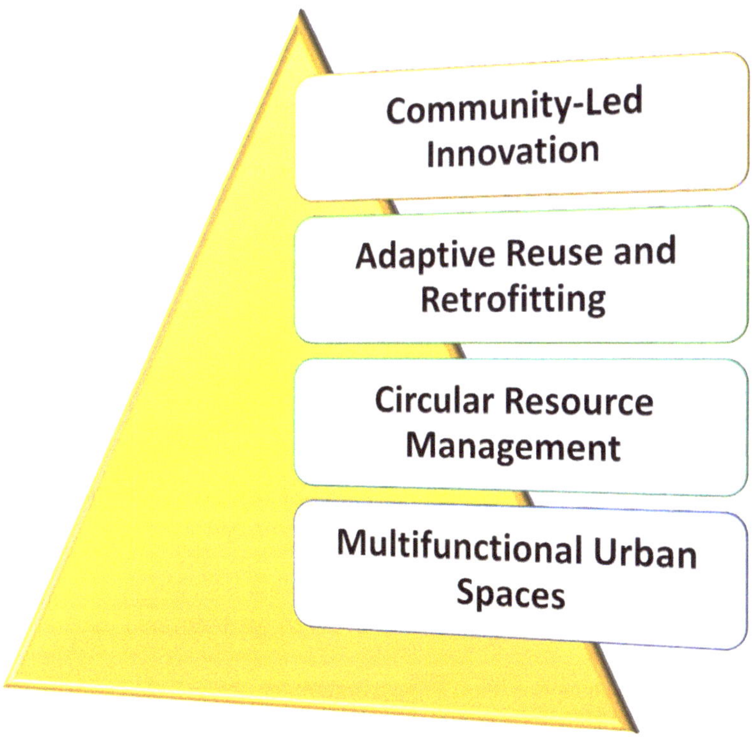

Fig. 1.1 Four pillars of the book

reader's presumptions about mainstream urban development and encourages them to interact with the case studies that follow—not as stand-alone experiments, but as indicators of a more profound paradigm shift that is already under progress.

1.4.1 Community-Led Innovation: The Democratic Engine of Resourceful Urbanism

The first thematic pillar emphasizes how communities' creative potential serves as the main force behind urban change. All too frequently, state-led megaprojects, corporate R&D, and labs are associated with innovation. However, the most resilient types of innovation are frequently driven by people who experience scarcity daily and arise out of necessity. Through community-led innovation, locals are reframed as active designers of their urban futures rather than passive recipients of planning benefits.

This book shows how social entrepreneurship, cooperative governance, and grassroots organizing create robust, contextually grounded infrastructures. Communities use scarce resources to create long-lasting homes, services, and public space, as seen in Fiji's Koroipita Model Towns and London's South Bank, which was revitalized by locals utilizing Coin Street Community Builders. These instances demonstrate how urban development is supported by collective activity rather than just capital.

By placing the community at the center of creativity, this theme makes the case that democratic involvement is the driving force behind creative urbanism rather than just a formality. New forms of design, finance, and government are created via grassroots movements, co-ops, and neighbourhood committees. Readers are encouraged to see how, even under situations of severe limitation, the democratic imagination may create infrastructures of justice, resilience, and belonging.

1.4.2 Adaptive Reuse and Retrofitting: The Art of Urban Archaeology and Renewal

The second pillar turns to the material fabric of the city. The history of labour, extraction, and circulation are embodied in every street, warehouse, and hallway. According to traditional development models, renewal starts with a blank slate, and obsolescence results in demolition. In contrast, resourceful urbanism views the current city as a rich archaeological site that has to be discovered, repurposed, and revitalized.

Retrofitting and adaptive reuse serve as examples of how heritage and embodied energy saving can be used to create new civic and cultural roles. A former freight rail route along Sydney's Goods Line serves as a pedestrian spine that connects neighbourhoods, museums, and institutions. Africa's largest museum of modern art, Zeitz MOCAA, is carved from an abandoned grain silo in Cape Town. Each example demonstrates how resourcefulness is found in creative transformation rather than erasure, repurposing ancient forms for new purposes and bringing the memory of a place into line with new goals.

This topic contends that the sustainable city of the future will be constructed using the foundations of the city we already have, challenging the craving for the new. Retrofitting is not second-best; rather, it is an art of renewal and a form of urban archaeology that reveals the space's multi-layered histories while building infrastructure that is appropriate for today's demands.

1.4.3 Circular Resource Management: The Metabolic Redesign of the City

If the first two pillars emphasise people and places, the third focuses on flows. Cities are metabolic systems that emit waste, emissions, and pollutants while consuming resources like food, energy, water, and materials. These flows are treated as linear by conventional models, which result in depletion at one end and accumulation at the other. This metabolism is rearranged into loops by circular resource management, which lessens reliance on extraction, aligns ecological cycles with urban economies, and reroutes waste as input.

The cases examined here range from Pune's SWaCH cooperative in India, where waste pickers turn informal recycling into a recognized public service, to the internationally recognized Kalundborg Industrial Symbiosis in Denmark, where companies exchange steam, water, and by-products in closed loops. These are living practices of metabolic redesign based on both high-level industrial cooperation and routine household collecting; they are not abstract blueprints for the circular economy.

The lesson is obvious: circularity encompasses social inclusion, governance, and trust in addition to technology advancements. Circular resource management shows how collaboration and creativity may rethink urban metabolism by enlisting waste pickers, municipal utilities, industrial partners, and households. This theme presents readers with a positive view of cities as regenerating metabolisms as well as a critique of linear extractivism.

1.4.4 Multifunctional Urban Spaces: The Spatial Synthesis of Efficiency and Delight

The areas where ecological, cultural, and social functions come together are highlighted by the fourth pillar. Land is frequently divided into single-use zones in conventional planning, such as residential, commercial, and recreational. The reverse is demonstrated by resourceful urbanism, which layers several uses into a single area to create both efficiency and enjoyment.

Zoorasia in Yokohama and the nearby Forest Park serve as examples, integrating zoological displays, ecological restoration, education, and enjoyment into a single, cohesive landscape system. They also include Medellín's Library Parks, which combine schools, plazas, libraries, and cultural programming in areas that were formerly characterized by isolation and violence. Multifunctionality increases value in both situations: a zoo becomes an ecological classroom; a park becomes a flood buffer; and a library becomes a municipal anchor.

Planners and designers are challenged by multifunctional urban areas to view land as a living canvas where social, ecological, and cultural functions coexist rather than as a limited resource to be distributed. This spatial synthesis, i.e., the creation of spaces that satisfy several demands while stimulating a fresh sense of urban imagination, is where the ingenuity lies.

1.5 A Call for Rethinking: Toward a Resourceful Revolution

When combined, these four themes serve as more than just a framework for this book. They mark a significant change in the conception, planning, and governance of cities. The following eight case studies—from Sydney to Medellín, from London to Lautoka—are not side projects. They portend a time when ecosystems, institutions, and communities work together to build regenerative cities and scarcity is not seen as a threat but rather as an opportunity.

For this reason, the book emphasizes on portraying Resourceful Urbanism as an ongoing revolution rather than a far-off dream. Communities everywhere are rejecting to wait for ideal circumstances or limitless funding. They are making use of their resources, which include social solidarity, industrial byproducts, abandoned land, informal labour, and cultural memory. They are building resilience, belonging, and regeneration infrastructures out of these basic ingredients.

The message is clear: you are already involved in this change, whether you are a planner choosing how to retrofit a district, a legislator creating legislation, an architect creating spaces, or a citizen demanding accountability. Participation, not spectatorship, is the goal of resourceful urbanization.

In addition to encountering creative case examples, readers will also gain a conceptual framework and useful actionable tactics by the end of this book. They will observe that communities and institutions all around the world are already implementing the seeds of regenerative futures in their daily operations. All that is left to do is identify, magnify, and link these practices into a cohesive movement.

The ensuing voyage is both instructive and thought-provoking: it maps inventive methods across countries and challenges us to reconsider what cities can become. In a time of ecological limits, inequality, and climate breakdown, creative urbanism provides promise in addition to critique. It invites us to reimagine the city as a site of ingenuity, solidarity, and regeneration, a place where abundance can be found within constraints.

References

1. Chenic AŞ, Cretu AI, Burlacu A, Moroianu N, Vîrjan D, Huru D, Stanef-Puica MR, Enachescu V (2022) Logical analysis on the strategy for a sustainable transition of the world to green energy—2050. Smart cities and villages coupled to renewable energy sources with low carbon footprint. Sustainability 14(14):8622
2. Enkvist P, Nauclér T, Riese J (2008) What countries can do about cutting carbon emissions. McK Quarterly 2:34
3. Cheshmehzangi A (2025) Urban planning for the contemporary age: navigating complexities and shaping urban futures. Encyclopedia 5(1):19
4. Rogan K (2019) Anti-intelligence: a Marxist critique of the smart city. Doctoral dissertation, Masters Thesis, Theories of Urban Practice, Parsons School of Design
5. Kostakis V, Pazaitis A, Liarokapis M (2023) Beyond high-tech versus low-tech: a tentative framework for sustainable urban data governance. Big Data & Society 10(1):20539517231180583
6. Allen J, James AD, Gamlen P (2007) Formal versus informal knowledge networks in R&D: a case study using social network analysis. R&D Manag 37(3):179–196
7. Cheshmehzangi A (2025) Stewarding the living fabric: natural environment as the Resource Base of advanced cities. In: Advanced City: for the future of cities and opposing smart cities. Springer Nature Singapore, Singapore, pp 215–225
8. Kibert CJ, Sendzimir J, Guy B (2000) Construction ecology and metabolism: natural system analogues for a sustainable built environment. Constr Manag Econ 18(8):903–916
9. Allam Z, Cheshmehzangi A (2024) Theoretical and international perspectives on sustainable futures. In: Sustainable futures and green new deals. Springer Nature Switzerland, Cham, pp 29–53
10. Verlinghieri E (2016) Planning for resourcefulness: exploring new frontiers for participatory transport planning theory and practice in Rio de Janeiro and L'Aquila. Doctoral dissertation, University of Leeds
11. Mouat C, Adams B (2023) Recognizing craft and creativity as political governance innovation activating people and place through civic activism and creative enterprise. Craft Communities 94
12. McNall SG (2015) The problem of social inequality: why it destroys democracy, threatens the planet, and what we can do about it. Routledge, New York
13. Wendler J (2014) Experimental urbanism: grassroots alternatives as spaces of learning and innovation in the city. Doctoral dissertation, University of Manchester
14. Agudelo-Vera CM, Mels AR, Keesman KJ, Rijnaarts HH (2011) Resource management as a key factor for sustainable urban planning. J Environ Manage 92(10):2295–2303
15. Healey P (2004) Creativity and urban governance. disP-The Plan Rev 40(158) 11–20
16. Scott AJ (2014) Beyond the creative city: cognitive–cultural capitalism and the new urbanism. Reg Stud 48(4):565–578

Part I
Community-led Innovation

Chapter 2
From Derelict Riverside to Residents' Commons: Coin Street Community Builders and the Politics of Collective Urban Stewardship

2.1 Context, Origins and Political Economy

The story of Coin Street starts in the tumultuous political and economic environment of London in the late twentieth century. The South Bank riverbank between Waterloo and Blackfriars was in a state of disrepair by the early 1980s. Once a thriving commercial and industrial district with warehouses, railroad sidings, and wharves, the region had been abandoned as industry decreased and port operations moved downstream. A strip of land that was both strategically positioned and socially peripheral was distinguished by abandoned plots, closed buildings, and parking lots [1]. Even though it was close to the South Bank Centre and the National Theatre, two cultural hotspots, the neighbourhood's problems were indicative of a larger trend of urban disinvestment: subpar housing, few job options, and a frayed sense of community.

At the same time, the Thatcher administration was changing the political economy of London. Frequently at odds with the national government, the Greater London Council (GLC) has been working on plans to safeguard public resources and promote more equitable growth [2]. In this contentious environment, state authorities and speculative developers viewed the South Bank as a possible location for massive commercial projects, such as office skyscrapers, hotels, and retail complexes, which were seen as a rapid path to capital accumulation and urban "renewal". However, these proposals threatened to privatize limited central-city land, displace many citizens, and destroy their sense of community.

Locals started to organize in this atmosphere of danger and opportunity with the help of organizers, GLC members, and empathetic activists. They characterized the issue as one of exclusion rather than a lack of private investment, since precious property was being used as a commodity to maximize profits rather than as a shared resource that might support social life, housing, and communal well-being [3]. This rephrasing was a significant change in discourse. In order to question the dominant

A. Cheshmehzangi, *Resourceful Urbanism*, Urban Sustainability,
https://doi.org/10.1007/978-981-95-4138-6_2

logic of market-led development and establish themselves as legitimate actors in determining the future of metropolitan areas, campaigners redefined derelict property as a chance for community ownership and management.

The establishment of Coin Street Community Builders (CSCB) in 1984 crystallized this response [4]. CSCB, which was set up as a social enterprise and development trust, represented the hybrid governance required to function in London's changing political environment. Its roots are in both grassroots activism and astute bargaining with official institutions. To build a barrier against speculative demand, locals took advantage of the GLC's readiness to transfer land and the newly developed instruments of cooperative housing and communal land ownership. By doing this, they illustrated how limitations—such as political animosity, little funds, and depreciated land—could be repurposed as advantages for innovation driven by the community. As a result, the Coin Street instance became a trailblazing example of resourceful urbanism: a product of hardship, characterized by struggle, and maintained by group efforts.

2.1.1 *Pre-intervention Context: South Bank in Late-twentieth-Century London*

By the late 1970s and early 1980s, London's South Bank between Waterloo and Blackfriars presented a striking paradox [5]. Geographically, it was situated just across from the financial area of the City of London, on one of the most accessible and central sections of the Thames riverbank. However, it was characterized by marginalization and neglect on a social and physical level. When river traffic moved downstream and industrial operations scattered to the metropolitan periphery in the post-war decades, the South Bank's economic foundation—once a center of wharves, warehouses, light industry, and railway depots—collapsed. Plots were turned up to surface parking lots, storage, and ad hoc uses, reflecting a lack of investment or strategic planning, while buildings remained empty or underutilized. The urban fabric of the area was broken up by arterial roads and train viaducts that separated neighbourhoods from the river, leaving the built environment scarred by abandonment.

This neglect took place when London's political economy underwent more significant changes. There was demand to repurpose strategically placed land for new uses as a result of the city's deindustrialization, especially for the commercial and financial industries that were growing quickly under Thatcher. The South Bank was increasingly viewed as a "frontier" for office and hotel developments by developers and real estate speculators [6]. At the same time, locals had to deal with worsening living conditions and the possibility of being displaced if extensive reconstruction went ahead unchecked. There was a clear conflict between community demands and market-driven ideas of revitalization.

Another level of intricacy was introduced by municipal politics. Under progressive leadership in the early 1980s, the Greater London Council (GLC) aimed to prioritize social goals in land management and oppose speculative proposals [7]. As a result, the council found itself at odds with both the federal government and private developers who preferred redevelopment based on profit. The councils of Lambeth and Southwark, which included portions of the South Bank, were also having financial difficulties and conflicting agendas [8]. Opportunities for several intervention models arose in this setting of disputed authority, dispersed land ownership, and conflicting future aspirations for the region.

This setting is vividly depicted in maps and aerial pictures from the time period: an underutilized area of land in the center of London that is cut off from the surrounding settlements and awaits a significant rethinking of its use. This environment of disregard, conjecture, and political conflict gave rise to the Coin Street movement.

2.1.2 *Emergence of the Resident Movement and Founding of Coin Street Community Builders (CSCB)*

The Coin Street story is best interpreted as the result of a combination of local willpower, shifting political allegiances, and market logic's inability to produce results that are in line with the welfare of the community. Residents of the South Bank were well aware of the strains affecting their neighbourhood by the early 1980s [9]. Developers were surrounding the riverfront lands, supported by a national policy climate that favoured deregulation and privatisation. Plans put up at the time called for hotel complexes and towers of office space in an effort to capitalize on the area's close proximity to important cultural institutions and the City of London. However, these ideas provided nothing for already-existing communities because they put profit ahead of social need, ran the risk of uprooting local households, and produced few truly accessible facilities.

Local tenant associations, resident activists, and empathetic planners who understood the dangers of unbridled speculative growth were important players in opposing this trajectory. An important ally was the GLC, which was run by Labour at the time and was more willing to try out community-driven development approaches. Residents' demands for safety and responsibility drew the Lambeth and Westminster councils, whose domains included the disputed sites, into the discussion.

The trigger events were several. Residents were enraged by the press's extensive coverage of public disclosures regarding speculative plans for the site in the early 1980s, which made them fear being excluded from choices that would change their neighbourhood. Community gatherings, which are frequently recorded in flyers and minutes from the time, became forums for planning, discussing options, and strengthening group commitment. To highlight the risks of market-led proposals, activists organized protests, petition drives, and media appearances. Early organizers' interviews show how the problem was clearly framed: it was about land ownership and whose interests urban development served, not only about buildings.

The idea to establish a community-controlled development trust arose from these conflicts. Established in 1984, CSCB symbolized the institutionalization of grassroots energy into an organization that could engage in negotiations with bankers, developers, and the government. The establishment of CSCB required both strategic pragmatism and political opportunity. Following discussions with the GLC, CSCB was granted favourable terms for the transfer of about 13 acres of South Bank land. The agreement demonstrated the movement's tenacity in presenting a different course as well as the council's intention to curb speculative excesses.

To comprehend why CSCB was feasible, it is essential to consider the failure of market-led solutions. Theoretically, commercial redevelopment might have generated income for the government and jobs. In reality, a gap was created by a lack of funding, shifting real estate markets, and a misalignment between speculative schemes and community needs. Developers' plans did not align with local or municipal interests beyond profit, and they were unable to provide sustainable returns without taking on a great deal of risk. An alternative driven by the residents was made possible by this disjunction.

2.2 Governance Architecture and Finance

A prime illustration of how creative finance and hybrid governance may support community-led urban renewal is CSCB [10]. Moreover, CSCB was purposefully designed from the start in 1984 to resist speculative influences and to place resident concerns at the center of decision-making. Its organizational structure represents a deliberate layering of institutional models: a network of housing cooperatives that permanently anchor inhabitants, a social enterprise arm to create revenue and reinvest surpluses, and a development trust to ensure long-term land ownership. A strong framework for striking a balance between community accountability and financial sustainability was established by this agreement.

At the heart of CSCB is its role as a development trust. Unlike private developers, the trust holds land and assets not for short-term profit but for long-term stewardship. By using the trust concept, CSCB was able to purchase from the Greater London Council about 13 acres of valuable South Bank land on favourable terms, protecting the property from speculative resale [4, 5]. The establishment of housing cooperatives, which operate as separate legal entities with their own membership, governance, and tenant decision-making authority, was a supplement to the trust system. With rights and obligations closely linked to their involvement in cooperative governance, residents essentially became shareholders in their homes. This co-op model integrated collective governance into the fundamental structure of housing provision while offering affordability and tenure security [11]. In addition to housing, CSCB set up commercial arms and subsidiaries, such as event venues and property management companies, to make sure the trust had a varied portfolio of income-producing ventures.

Financial sustainability has always been central to CSCB's survival. The organisation pursued a blended finance strategy that combined public grants, community fundraising, and private borrowing. Initial leverage was supplied by early seed money from the GLC and later assistance from benevolent municipal sources. After thereafter, these monies were carefully utilized to get private financing for building and remodelling. Crucially, CSCB incorporated the idea of cross-subsidization, whereby commercial rents from retail establishments, office buildings, and event spaces were used to directly fund community initiatives and social housing subsidies. As a result, market activity became a method of maintaining non-market outcomes rather than an end in and of itself, creating a positive feedback loop. A financial overview of CSCB's records shows a conscious balancing act, with surpluses being reinvested in new initiatives like community gardens, childcare centers, and cultural programming, with income from commercial tenants occasionally compensating for cyclical declines in grant support (see Table 2.1).

Participatory governance was the critical counterpart to financial innovation. In addition to experts with experience in planning, finance, and law, CSCB's board has always included local representatives. While tenant councils within housing cooperatives ensured bottom-up involvement on day-to-day management, annual general meetings provided a forum for residents to ask questions and set strategic direction. Accountability measures included required audits, transparent reporting,

Table 2.1 Governance and finance architecture of CSCB

Component	Core function	Examples/Mechanisms	Contribution to resourceful urbanism
Development trust	Long-term land stewardship	GLC land transfer (13 acres), prevention of speculative resale	Secures land as a commons rather than a commodity
Housing cooperatives	Resident-led governance and tenure security	Separate legal entities, tenant decision-making councils	Embeds democratic participation, affordability, and community accountability
Social Enterprise arms	Generate income and reinvest surpluses	Commercial rentals (retail, offices, event venues)	Creates cross-subsidy loop to sustain social and housing programs
Blended finance strategy	Financial resilience and risk management	Mix of public grants, concessional loans, community fundraising, private borrowing	Reduces reliance on single funding source, enables incremental development
Participatory governance	Transparency, accountability, inclusivity	Board with local reps + professionals, tenant councils, annual AGMs	Balances technical expertise with grassroots legitimacy
Risk mitigation	Protection against shocks	Diversified income, long-term leases, political alliances	Sustains continuity through recessions, political change, and shifting markets

and frequent newsletters that explained choices in plain terms. Trust between residents and the trust's leadership was preserved through the use of conflict resolution techniques, such as organized consultations and mediation procedures. However, critical evaluations point out drawbacks: although locals had significant representation, the governance system also necessitated professionalization, which occasionally stifled grassroots voices in favour of technical know-how. One of the key characteristics of CSCB's governance model is still the conflict between efficiency and inclusivity.

Risks and constraints have been ever-present. Since the GLC was abolished in 1986 and the municipal leadership in Lambeth and Westminster changed, CSCB has had to deal with changing political environments [12]. Financially speaking, dependence on commercial rentals made one vulnerable to market downturns, particularly during recessions or times when offices were vacant. Another difficulty was maintaining communal amenities and cooperative housing, as if long-term maintenance expenditures weren't carefully controlled, they would eventually surpass income. CSCB used diversity to mitigate these risks. It protected itself against volatility by diversifying its sources of revenue, obtaining long-term leases with reliable tenants, and preserving alliances with local government and cultural organizations. The trust also made an investment in qualified management skills to make sure it could react swiftly to financial or regulatory demands without losing sight of its goal.

Taken together, CSCB's governance and finance architecture exemplify resourceful urbanism in action. It illustrates how societies can create institutional frameworks that transform limitations—such as scarce capital, political animosity, and speculative pressure—into long-lasting organizational structures. Land is held in community hands by the development trust, social equality is ingrained by the cooperatives, and long-term sustainability is maintained by the social enterprise arms. This hybridity, which maintains that urban land is a common to be managed collaboratively rather than a commodity to be exchanged, is not just technical; it is philosophical. The approach has had drawbacks, especially the possibility of creeping commercialization, but its longevity over many years is evidence of the strength of governance frameworks that strike a balance between fiscal pragmatism and community accountability. The takeaway from analysing CSCB's design is straightforward: organizations that are as innovative as the communities they serve are necessary for sustained urban development.

2.3 Implementation, Urban Design and Programmatic Outcomes

Under CSCB, the South Bank underwent a well-planned approach that prioritized experimentation, incrementalism, and community learning over extensive, expensive redevelopment. To reduce financial risk and make sure that every stage strengthened community trust in the project, this staged development approach was essential. In 1984, CSCB started a phased program of interventions after transferring about 13 acres of land from the Greater London Council. The initial stages concentrated on stabilizing and reusing pre-existing buildings, such as turning abandoned warehouses into makeshift offices and gathering places.

These quick-win, low-cost modifications gave the trust an instant presence, communicated real advantages to the inhabitants, and produced a small amount of rental revenue. As funds became available, new-build housing cooperatives, childcare centers, and community centers were gradually added in later phases. Through strategic sequencing, CSCB was able to steer clear of excessive leverage, modify designs in response to community input, and prove viability to doubtful funders and partners.

These interventions were guided by design ideas that reflected the spirit of resourceful urbanism [13]. A recurring theme was adaptive reuse, whereby pre-existing structures were not only demolished but instead repurposed for a variety of uses, including offices, studios, and cultural and educational facilities. Where new building was done, it was carefully woven into the existing urban fabric, giving mixed-use layouts priority and preserving a human scale. The most obvious example are the housing cooperatives. Located around common courtyards that strengthened social life, low- to mid-rise buildings provided a varied mix of inhabitants with secure, reasonably priced housing. Floor plans promoted social contact without compromising privacy by striking a balance between private residential space and communal features like gardens and common areas. The placement of community centers and halls created porous boundaries between the co-ops and the city, catering to both locals and the general South Bank population. What had previously been disparate land parcels were brought together by outdoor spaces such as playgrounds, gardens, and groomed pathways. The focus on multipurpose areas, green margins, and walkability made sure that design interventions were socially and environmentally inclusive.

Programming and operations were equally central to CSCB's success. The physical interventions were activated through a diverse array of activities that nurtured social capital and local economies. By providing a location for childcare services, after-school programs, and adult education, the Coin Street Neighbourhood Center addressed urgent social needs and supported parents' employment potential. Weekend markets strengthened the South Bank's reputation as a vibrant civic area by bringing life to public areas and giving local business owners, craftspeople, and food vendors a stage. From exhibitions to performances, the arts programming made the most of the location's close proximity to important cultural institutions while guaranteeing that locals, not just visitors, could enjoy the culture. These programs promoted a distinct form of cross-subsidy: just as commercial rents maintained financial stability, cultural vibrancy and livelihood chances maintained community identity and economic resilience.

Long-term placekeeping, stewardship, and maintenance were not afterthoughts; rather, they were incorporated into the project's governance and design [14]. With the help of service fees and voluntary contributions from residents, housing cooperatives took up a large portion of the regular maintenance of common areas. The management of larger buildings and public spaces was overseen by CSCB's qualified staff, who made sure that health and safety laws were followed while preserving aesthetic standards. Crucially, income from commercial rentals and event space was set aside for continuing maintenance as well as debt repayment and new construction, avoiding the deferred-maintenance trap that threatens a lot of community initiatives. A further layer of stewardship was added by volunteer-led projects, like community gardening clubs and events planned by residents, which strengthened the sense of place and ownership among the locals.

Both quantitative and qualitative results are evident from this multifaceted strategy. Strong resident satisfaction with housing quality, tenure security, and the availability of community facilities was found in surveys carried out in the 1990s and 2000s. Market and event attendance often surpassed projections, and the number of annual community events increased rapidly. Maps showing the South Bank before and after show the dramatic change from an archipelago of abandoned warehouses and parking lots to a unified metropolitan area thriving with housing, business, and culture. From weekend cultural festivals to weekday daycare services, seasonal activity heatmaps would show a constant rhythm of use, demonstrating how rich, continuous programming has magnified small physical spaces.

The integration of physical design, incremental delivery, and daily programming is what sets CSCB's execution apart, not the majesty of its architecture. By placing a high value on affordability, flexibility, and group responsibility, CSCB established an environment where ingenuity is ingrained in the very fabric of the city. Today's South Bank serves as proof that urban revitalization does not always require a high-capital spectacle; rather, it may be fuelled by small-scale initiatives that, when maintained by organizational tenacity and community imagination, can result in significant change. The core of resourceful urbanism is a concept that fosters a durable, living commons through the reuse of devalued land and mutually reinforcing design and program.

2.4 Measured Impact, Externalities and Critical Appraisal

CSCB's impact must be evaluated using both qualitative and quantitative methods. The project's results go beyond the quantity of housing units constructed or the square meters of public space provided; they also contain less obvious but no less significant contributions to urban identity, government resilience, and community cohesiveness. However, a thorough evaluation must also address criticisms of unforeseen repercussions, such as the threat of gentrification pressures and the potential for community projects to be commodified. According to the research, CSCB has made significant but challenging progress, and its resilience is attributed to its capacity to continuously adjust to changing social, political, and economic environments.

In terms of housing and economic results, CSCB has produced a stable and observable supply of affordable housing in one of the most strained regions of

London. A variety of households who would otherwise be priced out of the core city now have secure tenure thanks to the trust's development of more than 200 cooperative housing units spread over many sites by the early 2000s. Because these units were purposefully designed using cooperative concepts, inhabitants had rights and obligations as group stakeholders. In an area that would otherwise be dominated by opulent apartments and speculative office complexes, the affordability of these residences has been essential in maintaining a varied resident base. In addition to housing, CSCB created about 50,000 square feet of retail and commercial space, which is leased to a variety of small enterprises, cafés, and cultural organizations [4, 5]. In addition to directly generating local job possibilities, this commercial activity produced cross-subsidized income to support social programs and upkeep. According to estimates from CSCB's records, its operations directly supported hundreds of employments, including those of daycare providers, retail employees, and maintenance workers. Tens of thousands of square feet of community and commercial floor space, about 220 affordable units, and consistent cash streams that are reinvested in the local economy would all be included in a hypothetical table of essential criteria.

The project's social and civic effects go far beyond its physical attributes. Living in housing maintained by CSCB is regularly linked to a strong sense of security and belonging, according to resident surveys and interviews. Because members actively participate in decision-making rather than being passive beneficiaries of housing, the cooperative model has been especially successful in fostering democratic involvement. Even though it takes a lot of effort, many residents find the process empowering, and reports from annual general assemblies and tenant meetings indicate high participation rates. Another advantage has been diversity: unlike the homogeneity of neighbouring private complexes, CSCB housing supports a variety of ethnicities, economic levels, and family configurations [4]. CSCB's community programs, such as childcare, adult education, and youth participation, have had quantifiable impacts on social capital outside of the residential setting. For instance, more than 100 families attended the neighbourhood center's childcare services each year, while dozens of students attended its adult education classes each term. Researchers gathered interviews that show young people gaining from cultural programming, older residents reestablishing links with the community, and parents able to resume their jobs. These anecdotes illustrate how CSCB's activities improved regular civic life and serve as a supplement to the quantitative statistics.

Urban and environmental outcomes are equally visible. Prior to CSCB's interventions, the South Bank site was characterised by vacant lots and derelict warehouses. The neighbourhood is now a thriving urban quarter with lively public areas that hold seasonal festivals, performances, and markets. Weekend markets are a mainstay of the local economic and cultural life, and outdoor events frequently draw thousands of visitors. In addition to improving the riverfront's environmental quality, the green spaces developed by CSCB have somewhat raised the area's per capita access to open space when compared to suburban norms. The adaptive reuse of existing structures and the reduction of vacancy itself have environmental value, even though CSCB has not explicitly prioritized environmental sustainability like other urban projects have. This is because the project avoided the embodied carbon costs of wholesale demolition and speculative redevelopment by prolonging the life of buildings and land.

Critiques and unforeseen consequences, however, merit consideration. Ironically, CSCB's success has aided in the South Bank's broader gentrification. Higher-end constructions have been drawn to the region due to its vitality and the enhanced public realm, which has increased property values and created demands that go beyond the limits of CSCB-owned land. Although CSCB has maintained affordability within its holdings, critics contend that it has failed to protect neighbouring communities from demands to relocate. Another criticism focuses on the possible commercialization of community projects. According to some, CSCB runs the risk of losing its grassroots identity and functioning more like a developer than a movement now that it is a recognized and professionalized organization. Funding for maintenance has also been difficult. Even though commercial rents have historically kept businesses afloat, vacancies or downturns in the commercial real estate market present dangers, underscoring the shortcoming of cross-subsidization as a long-term strategy. Last but not least, political developments like the 1986 repeal of the GLC and the reorientation of local government goals have compelled CSCB to continually renegotiate its legitimacy, highlighting both its adaptability and its susceptibility to more significant structural changes.

But what really sticks out is how strong and long-lasting CSCB's governance is. The company has survived several market cycles, political upheavals, and demographic changes over the course of almost 40 years. It has been able to adapt without losing sight of its primary goal because to its governance systems, which combine professional knowledge with resident representation. Consistent personnel and well-defined organizational procedures have preserved institutional memory, and board succession has been controlled to maintain continuity while introducing fresh viewpoints. Its hybridity—being a community organization, housing provider, landlord, and cultural operator all at once—is what gives CSCB its longevity. Because of its diversity, it has been protected against single points of failure and has the flexibility to change course when necessary.

From a counterfactual standpoint, the South Bank would most likely have produced exclusive luxury homes and office space with little amenities for nearby neighbourhoods if speculative commercial development had predominated there in the 1980s. CSCB showed that creative urbanism may provide long-term affordability, social capital, and civic vibrancy by leveraging the opportunity presented by political will and market failure [15]. Its accomplishments, albeit not without contradictions, are quantifiable and noteworthy. It continues to serve as a reminder of the limitations of local efforts in the face of structural urban inequality as well as a paradigm of what community-led urban development can achieve. The reason CSCB is such a remarkable case is because of this duality: it invites continuous critical examination while providing proof of possibilities.

2.5 Lessons, Transferability and Policy Recommendations

For practitioners, legislators, and communities looking to reinvent urban development through innovative and inclusive approaches, the Coin Street case provides a multitude of lessons. Its accomplishments are based on a meticulous layering of governance, finance, design, and programming that turned limitations into possibilities rather than on a wealth of capital or state-led master planning. This section breaks down the key lessons, examines how they might be applied in different situations, and lists the institutional resources and legislative tools that could help comparable models spread.

A collection of practical lessons with widespread applicability lies at the heart of Coin Street's success. First, mission longevity is underwritten by hybrid funding. CSCB protected itself against market or political volatility by reducing its reliance on any one source of funding by blending grants, loans, community investment, and commercial rents. Second, phased development and incremental acquisition lower financial risk and political tension. Instead of delivering everything at once, CSCB showed how small, initial successes, like renovated warehouses and community programs, might generate momentum and credibility for more significant actions. Third, space is turned into a source of income through embedded programming. Buildings were not merely shells; they were lived infrastructures that produced everyday social and economic value thanks to housing, childcare, marketplaces, and cultural events. Fourth, accountability is maintained through mixed government. By combining community involvement with professional knowledge, systems were developed that were both locally legitimate and robust enough to resist outside influences.

Lastly, by integrating stewardship into the very fabric of development, institutional resilience is achieved by presenting land as a commons rather than a commodity. Differentiating which elements of the Coin Street model are transferable and which are dependent on particular local circumstances is essential when thinking about transferability. The following table illustrates how feasibility is shaped by important enabling factors (see Table 2.2).

The takeaway is that the Coin Street approach is neither strictly unique nor globally reproducible. It flourishes when underappreciated urban land, receptive political allies, and active communities come together. Partial transfer is feasible even in less favourable circumstances, though, as techniques at all scales—from neighbourhood projects to urban redevelopment—can be informed by incrementalism, hybrid finance, and embedded programming.

Replication requires institutional support and policy levers. Municipal laws that permit cooperative housing or community land trusts are essential because they establish legal frameworks for common ownership that are resistant to speculative pressures. By proactively acquiring and holding onto abandoned land for future transfer to community trusts, land banking policies can help cities avoid premature speculative capture. Fiscal instruments that enable community-led organizations to get initial traction include social impact bonds, revolving community lending funds,

Table 2.2 Summary of key findings based on six key dimensions of the project

Dimension	High transferability	Context-dependent	Low transferability
Land availability	Rare urban dereliction or undervalued sites are essential; more feasible in post-industrial zones than hyper-saturated land markets	✓	
Municipal politics	Supportive local authorities and enabling legislation critical; oppositional regimes may block transfer	✓	✓
Access to seed finance	Moderate transferability; requires catalytic grants or concessional loans to establish trust	✓	
Civil society capacity	Essential; without organised resident groups, replication falters	✓	
Cooperative housing culture	Easier in contexts with traditions of co-op tenure (Nordic, UK, Canada); harder in regions dominated by speculative private ownership	✓	✓
Cross-subsidy via commerce	Strong in mixed-use districts with demand for retail/office; weaker in small towns or declining markets	✓	

and seed grants. Development frameworks can be skewed toward social outcomes by planning tools like mixed-use integration requirements, zoning benefits for cooperative housing, or preferred use covenants [16]. Equally important is investment in civil society capacity: training programmes, technical assistance, and knowledge-sharing networks help resident groups develop the organisational sophistication needed to manage complex projects.

The Coin Street incident also emphasizes how governmental actors must function more as catalysts than as controllers. An important example of how governments can empower people by sharing control rather than monopolizing it was the GLC's decision to transfer land on favourable terms. Policymakers of today ought to think about the ways in which regeneration techniques, surplus public assets, and land value capture might be used to the advantage of the community. Finding a balance between institutionalization and grassroots vitality is difficult because too much formalization can lead to alienation, while too little formalization can lead to fragility.

In conclusion, Coin Street demonstrates that, with the right institutional and financial support, community-led redevelopment is a feasible approach rather than a pipe dream. A realistic objective is suggested by the lessons learned from its experience: diversify funding, build gradually, provide extensive programming, involve the community in governance, and reframe land as a commons. Other cities can modify this approach to fit their unique needs by utilizing the proper policy levers, such as fiscal innovation, community land trusts, land banking, supporting planning frameworks, and capacity-building [17]. Although the transfer is not inevitable, inclusive, resilient, and regenerative urban futures elsewhere can be influenced by the resourceful urbanism demonstrated on London's South Bank with enough imagination and political will.

2.5.1 *Final Reflections: Reframing Constraints as Design Brief*

In the end, the Coin Street experiment shows how limitations—political animosity, financial scarcity, depreciated land, and fragmented governance—can be reframed as opportunities for innovation rather than obstacles. Residents completely reframed the issue, whereas traditional developers viewed an abandoned riverbank site as a chance for financial benefit. They saw the problem as "converting exclusion into stewardship and scarcity into solidarity," rather than "how to attract capital". The core of resourceful urbanism is this reinterpretation. To create long-lasting public benefit, it moves the focus of design away from the manipulation of form and money alone and toward the mobilization of institutions, connections, and modest resources.

The theoretical stakes here are significant. In mainstream development paradigms, the city is often reduced to a canvas on which capital is deployed to produce measurable returns. Housing is commodified, infrastructure is built as a means of promoting market competitiveness, and land is valued primarily for its trade potential. Under such a framework, design issues are resolved through significant renovation, iconic architecture, or financial infusions. This reasoning was reversed by Coin Street. Its design brief was determined by the need to harness collective capacity rather than the availability of funding. Social networks, legal resources, cooperative structures, resident expertise, and a readiness to engage in strategic negotiations with institutions were the "materials" at hand. The end result was a well designed urban quarter that continues to foster affordability, cultural activity, and civic pride rather than a spectacular architectural statement.

Coin Street is more than just a local success story when viewed in this light. It is a methodological intervention: a reminder that design encompasses institutions and governance in addition to buildings and environments. Resourceful urbanism views land as a disputed commons that needs to be managed rather than as a passive substrate for growth. It recognizes that social processes, such as meetings, discussions, and cooperative assemblies, are also acts of design, creating the frameworks that influence civic life and constructed form. This broadens the scope of the designer, planner, and legislator, requiring consideration of both the aesthetics of form and the choreography of relationships.

There are two wider ramifications. First, according to Coin Street, the issue of who owns land and for what reason should be the first step in any sustainable urban renewal process. Even the most well-meaning initiatives run the risk of getting caught up in cycles of speculation and exclusion in the absence of communal governance [18]. Secondly, it illustrates that incrementalism is not always a bad thing. A more capital-intensive "all at once" rehabilitation might never have created the resilience that CSCB did by constructing gradually, with each stage bolstering community legitimacy. This lesson is applicable far beyond London: resourcefulness can be the deciding factor in providing long-lasting public benefit in situations when financial resources are limited but social potential is vast.

But it's important to remember that Coin Street is vulnerable to more powerful structural pressures. Pressure is still being generated by the gentrification of the nearby South Bank and the increase in land values. Instead of becoming a paradigm shift for the entire city, the project continues to be a protective haven. This restriction, however, emphasizes how crucial it is to scale creative strategies with institutional support and policy frameworks. The resilience of CSCB implies that community-led models can endure in harsh settings with the correct scaffolding.

Therefore, Coin Street's legacy is seen in the way it reframes the design challenge [5]. Resourceful urbanism asks how to turn hardship into advantage, scarcity into solidarity, and exclusion into empowerment [19], in contrast to conventional models that ask how to allocate investment. Not all cities can recreate Coin Street perfectly, but they can all learn to view its limitations as a guide to be followed rather than challenges to be surmounted.

Successful urban futures will not be guaranteed by ever-increasing cash flows or outstanding architectural icons, as this book makes clear throughout. They will arise from ingenious methods that energize social forces, integrate land into communal care, and turn small-scale initiatives into long-lasting value systems. Coin Street Community Builders serves as a stark reminder that communities' inventiveness, which reframes limitation as possibilities, is what creates cities rather than just cash. The seeds of inclusive and regenerative urbanization germinate during that act of rephrasing.

References

1. Haworth G (2013) Coin Street Housing: the architecture of engagement. In: Sustainable urban design. Taylor & Francis, pp 117–131
2. Street E (2014) Sustainable governance and planning in London. In: Sustainable London?. Policy Press, pp 67–90
3. Tuckett I (1988) Coin Street: there is another way... Community Dev J, 23(4), 249–257
4. Brooke S (2024) London, 1984: conflict and change in the Radical City. Oxford University Press
5. Young K (1984) Governing Greater London: the background to GLC abolition and an alternative approach. London J 10(1):69–79
6. Sood A (2021) The speculative frontier: real estate, governance and occupancy on the metropolitan periphery. South Asia Multidiscip Acad J (26)
7. Allen J (1994) Consensus-Building in Metropolitan Planning: A Case Study of the Greater London Council, 1981–1986. Available from https://escholarship.org/content/qt8sm1z62s/qt8sm1z62s.pdf
8. Tyler D, Morad M (2008) London's South Bank'[1]. City Spaces-Tourist Places: Urban Tourism Precincts, 295
9. Baeten G (2001) Urban regeneration, social exclusion and shifting power geometries on the South Bank, London. Geogr Z:104–113

10. Tan SS (1998) The community as a partner in urban regeneration. University of London, University College London (United Kingdom)
11. Cheshmehzangi A, Munday R (2022) From global to local: the case of migration and urban identity for regenerative city transformations. J Urban Regen Renew 15(3):308–324
12. Baeten G (2000) From community planning to partnership planning. Urban regeneration and shifting power geometries on the South Bank, London. GeoJournal 51(4):293–300
13. Cheshmehzangi A (2025) Building industry infrastructure: foundations for sustainable, resilient, and integrated urban development. In: Advanced City: for the future of cities and opposing smart cities. Springer Nature Singapore, Singapore, pp 393–403
14. Burton M, Dempsey N, Mathers A (2014) Connecting making and keeping: design and management in place-keeping. In: Place-keeping. Routledge, New York and London, pp 125–150
15. Birolini J (2014) Towards a new form of economic development: why the need for an ownership revolution has never been greater. Available from https://base.socioeco.org/docs/dissertation_-_jerome_birolini_-_august_2014.pdf
16. Moos M, Vinodrai T, Revington N, Seasons M (2018) Planning for mixed use: affordable for whom? J Am Plan Assoc 84(1):7–20
17. Lall SV (2013) Planning, connecting, and financing cities—now: priorities for city leaders. World Bank Publications, Washington, DC
18. Cheshmehzangi A (2025) The Advanced City model: opposing the Smart City delusion and reclaiming 'the City we need'. In: Advanced City: for the future of cities and opposing smart cities. Springer Nature Singapore, Singapore, pp 467–479
19. Ziafati Bafarasat A, Oliveira E (2023) Social sustainability: do-it-yourself urbanism, start-it-yourself urbanism. Geoforum 141:103726

Chapter 3
Building with Scarcity: Koroipita and the Resourceful Urbanism of the Pacific

3.1 Urbanisation, Vulnerability, and the Political Economy of Housing in Fiji

It is impossible to separate Koroipita's story from Fiji's larger urbanization and vulnerability dynamics. Fiji has seen a consistent rural-urban migration over the last forty years as households have moved to towns like Lautoka, Suva, and Nadi in search of work. As the nation's second-biggest city and a major industrial center, Lautoka in particular attracted rural migrants looking for wage work in the port, tourism, and sugar sector. However, the rate of urban growth exceeded the ability of central government and local authorities to supply safe, affordable housing. As a result, families built temporary dwellings on marginal areas without stable access to amenities or tenure security, leading to the growth of informal settlements on the outskirts of cities [1]. In Fiji, an estimated one in five urban dwellers lived in such informal conditions by the early 2000s [2]. This statistic is similar to that of other Pacific Island countries where cities were finding it difficult to keep up with the rapid population growth due to limited economies and land constraints.

The vulnerabilities of these settlements were stark. Houses were particularly vulnerable to cyclonic storms, landslides, and flooding because they were situated on low-lying or steeply sloping terrain. The materials used for construction were frequently tarpaulins, corrugated iron, and scavenged timber, which provided little protection from storms. Every cyclone season carried the danger of destruction, and when catastrophes did occur, as they often did, locals had to deal with cycles of devastation and reconstruction without much outside assistance [3]. Equally important were the social repercussions: overcrowding, inadequate sanitation, and precarious housing led to social marginalization, interrupted schooling, and surroundings of chronic health risk. For governments already burdened

A. Cheshmehzangi, *Resourceful Urbanism*, Urban Sustainability,
https://doi.org/10.1007/978-981-95-4138-6_3

with inadequate financial resources, the magnitude of the problem seemed insurmountable.

The structural vulnerabilities of the Pacific intensified these challenges. Like its neighbours, Fiji is very vulnerable to climate risks, has a limited economic base, and is exposed to changes in global commodity prices. As regular components of the climate system, cyclones are not uncommon occurrences. Cyclones Winston (2016), Ami (2003), and Kina (1993) all left behind destructive paths that disproportionately affected informal communities [4]. The state's capacity to implement extensive housing programs was limited by budgetary constraints. Fiji was unable to rely on large welfare programs or hefty subsidies like wealthier countries might. Effective upgrading or relocation policies were also hampered by bureaucratic lethargy and a lack of state capacity. In the aftermath of disasters, international help frequently poured in, but it was focused more on immediate relief than long-term change.

Traditional housing models did not work well in this situation. Relying on private developers, market-led provision was unable to provide affordable housing on the necessary scale. Low-income households were left out of the private sector's focus on middle- and upper-income markets. In contrast, state-led models were small-scale and episodic, hindered by fluctuating political objectives and financial limitations. There were few public housing estates, and they frequently fell into disrepair, and relocation plans were unsuccessful due to complicated land tenure issues and people' reluctance to relocate far from their places of employment. The disparity between supply and demand grew rapidly, and by the year 2000, it was clear that traditional methods were failing.

This environment of fragility, shortage, and policy failure is exactly where the concept of Koroipita emerged [5]. The project transformed the paucity of resources into a design brief rather than a barrier by redefining restriction as opportunity. Communities may mobilize sweat equity if the government was unable to afford to build homes. Modular architecture and inventive recycling may lower prices if long-lasting housing materials were pricey. Leasehold agreements and community governance might offer stability without allowing speculation if market processes were unable to provide tenure security. Furthermore, homes would be built to resist cyclonic winds and rains rather than crumble under them if frequent calamities were unavoidable.

The founders of Koroipita, particularly Peter Drysdale and the Lautoka Rotary Club, understood that creating a model village needed more than just actual homes—it also needed institutions that could support them [6]. This way of presenting exposure and scarcity as a design problem is a perfect example of resourceful urbanism's philosophy, which emphasizes using available materials, enlisting the help of social networks, and integrating resilience into the very fabric of a community. By doing this, Koroipita addressed the structural issues of Pacific urbanism as well as the housing shortage, providing an example of how communities might rethink their futures while facing limitations.

3.1.1 Pre-intervention Conditions: Informal Housing and Vulnerability in Lautoka

The urban outskirts of Lautoka painted a familiar but stark picture of Pacific informality prior to the construction of the first home at Koroipita in 2003. The city, which has over 70,000 residents and is the second largest in Fiji, was growing rapidly due to the combined effects of rural-urban migration and industrial jobs [7]. Families came seeking work in the port economy, the sugar industry, or as casual labourers, but there was a lack of reasonably priced accommodation. The majority of new newcomers settled in unplanned peri-urban clusters on marginal land because the state was unable to provide suitable shelter and the private market was geared toward higher-income households. Thousands of Lautoka's poorest citizens moved into these informal settlements, creating a patchwork of improvised homes around the city's eastern and northern edges.

Large, frequently multigenerational families that had average family sizes larger than those in formal housing areas were a defining demographic of these communities. As a reflection of the wider diversity of Fiji's population, the migrants included both indigenous iTaukei families and Indo-Fijian households [8]. Low and erratic household earnings were largely reliant on seasonal labour or unofficial trading. While teenagers had high rates of underemployment, women frequently carried the twin responsibility of earning a living and taking care of the home. Settlements had a high population density, with several buildings and common amenities spread over tiny areas.

Systemic neglect was evident in the physical state of these settlements. Typically, houses were built using low-quality or salvaged materials, such as plastic sheeting, timber offcuts, and corrugated iron, and were put together gradually as families accumulated finances. Because so few buildings were anchored to foundations, they were extremely susceptible to cyclonic winds. Water and sanitary facilities were at best sporadic. Pit latrines or crude septic tanks were common, and communal taps or unofficial connections to municipal pipes supplied several families. At the fringes of settlements, open dumping was prevalent, and waste disposal was mainly unmanaged [4]. Connections to electricity were frequently shaky and informal. Tenure was precarious: a large number of households either squatted on state land, encroached on private holdings, or made ad hoc agreements with local landowners to occupy land without official lease agreements. Families were constantly at risk of eviction because of this uncertainty, which also deterred investment in long-term housing.

These risks were exacerbated by frequent floods and cyclones. Lautoka is located in Fiji's western division, which is frequently hit by tropical storms. In informal settlements, even mild cyclones caused havoc, causing roofs to be blown off, lightweight homes to collapse, and floods to wash through unprotected flooring [9]. For instance, reports from 2003's Cyclone Ami detailed extensive damage to Lautoka's squatter communities, leaving families without a place to live and dependent on temporary assistance. Recovery was sluggish and frequently involved reusing what was left to construct brittle dwellings in the same dangerous areas [1]. A cycle of vulnerability was strengthened by each disaster: losing possessions, having one's livelihood disrupted, and becoming even poorer.

The physical risks were exacerbated by social obstacles. Access to education was restricted by both cost and distance. Children from informal settlements faced difficulties with tuition, transportation, and repeated displacement during storms, despite the fact that Lautoka itself possessed schools. Dropout rates were greater than in formal neighbourhoods, and attendance was irregular. There were numerous health hazards. Malnutrition was exacerbated by inadequate diets and inconsistent incomes, while waterborne disease and respiratory infection outbreaks were made easier by poor sanitation and overcrowding [10]. Due to the distance or the lack of means for treatment and transportation, residents frequently did not have access to health care.

Opportunities for the economy were similarly limited. Residents of informal settlements faced discrimination in the job market and had limited access to permanent jobs, making them socially and spatially marginalized [1]. The majority of households rely on remittances from family members overseas or on casual labour, such as small-scale fishing, market retailing, or cane cutting. Families found it nearly impossible to cover the upfront expenditures of obtaining legal tenure or to obtain formal home finance in the absence of consistent income. Due to their limited options, inhabitants were obliged to continue living in subpar housing, which furthered the cycle of informality.

The planned, robust community that would eventually develop at Koroipita stands in sharp contrast to the visual evidence from this era [11]. Unserviced tracks and drainage ditches are scattered among haphazard clusters of tin-roof shacks in aerial photos of Lautoka's periphery. Images of cyclone destruction depict walls collapsing, roofs peeling apart, and families gathering supplies after storms. Marginal siting—settlements situated on steep, erosion-prone slopes or low-lying ground close to drainage channels—is highlighted on simple maps of pre-Koroipita land use. The environmental and geographical precarity of life on the outskirts of the city is highlighted by these images.

In short, stacked vulnerabilities—insecure tenure, precarious housing, insufficient services, and ongoing disaster exposure—defined the pre-intervention conditions un Lautoka's informal settlements. In addition to causing a housing shortage, these circumstances also fostered a more general atmosphere of social exclusion in which livelihoods, health, and education were continuously compromised. The establishment of a model town such as Koroipita was both necessary and feasible because of this environment, where the state and market systems had failed.

3.2 Emergence of the Initiative and Founding of Koroipita

The creation of Koroipita was neither the product of a grand state plan nor a market-led housing venture. Instead, it was the outcome of a very personal vision that was maintained through grassroots involvement and sparked by regional institutions [1].

The protagonist of this tale is Peter Drysdale, a businessman who was born in Australia and has lived in Fiji for a long time [12]. After personally observing the appalling conditions in Lautoka's informal settlements, Drysdale came to believe that affordable, long-lasting housing might be provided even in the face of acute shortage [6, 12]. His belief that improved housing would lessen disaster losses, increase family stability, and open doors to education and work was both pragmatic and moral. Drysdale turned this idea into a tangible project with the help of the Lautoka Rotary Club [6]. Credibility, small beginning finance, and important ties to international and governmental contributors were all made possible via Rotary's networks [6].

The project's driving principle was very different from traditional housing programs from the beginning. Drysdale had envisioned a "model town" that combined housing with social infrastructure, governance structures, and upward mobility pathways, as opposed to offering discrete units or one-time relocation operations [6]. Housing would withstand cyclones, but it wouldn't be isolated; the settlement's government committees, schools, community centers, and trash management would all be integrated into it [6]. The focus was on developing a cohesive and functional community that could support itself after the first construction phase, rather than just providing shelter.

Early partnerships were instrumental. Despite having little resources, the Fijian government granted regulatory clearances and later assisted with infrastructure connections. NGOs, such as Habitat for Humanity, provided technical assistance in community mobilization and low-cost building design. Financial support was provided by international philanthropists and development organizations that saw the project's potential as a replicable model for the Pacific. In addition to providing the framework for Drysdale's vision, these collaborations demonstrated the project's credibility outside of a single charismatic leader [6].

In 2003, the initial stage of development exemplified the sweat equity principle. After being chosen for relocation to Koroipita, families had to work to build both their own and their neighbours' homes [13]. This strategy not only lowered building expenses but also gave locals a feeling of pride and ownership. Families acquired building and maintenance skills that would eventually benefit the community by working with volunteers and knowledgeable craftspeople. Simple but sturdy, the initial group of homes served as evidence that affordable, cyclone-resistant housing could be achieved through teamwork.

The necessity of institutional government became evident as the colony expanded. The creation of the People's Community Network (PCN) transformed what started out as a Rotary-backed project into a more formalized framework [14]. This organisation provided a platform for residents to participate in governance, ensuring that decisions about maintenance, fees, and community programmes were made collectively. Additionally, PCN acted as a vehicle for advocacy, speaking on behalf of Koroipita people' interests before funders and the government. In order to transform Koroipita from a pilot project into a viable community with growth potential [1], the institutionalization process was essential.

Around 1000 people were housed in the second phase, Koroipita Stage 2, which increased the number of residences to over 230 by the middle of the 2010s [1, 6]. The success of sweat equity, the longevity of the cyclone-resilient design, and the significance of integrating education and youth programs early in the project were among the lessons learnt in Stage 1 that made this scaling feasible. Stage 2 expanded the involvement of resident committees in day-to-day management by adding more common infrastructure and improving the governance model.

Thus, the development and establishment of Koroipita demonstrate the interaction of institutional collaborations, community involvement, and visionary leadership. Its philosophy—that poverty, fragility, and disaster exposure are not insurmountable barriers but rather circumstances that can be intentionally used to build resilient communities—is what makes it unique. Koroipita went beyond the constraints of traditional housing schemes and demonstrated an innovative and reproducible model of urbanism by integrating social infrastructure [15] and governance into the project's fundamental fabric (See Table 3.1 for coalition map details).

Table 3.1 From Vision to Village: Coalition Map for Koroipita

Actor/Institution	Role in project	Core incentive	Key contribution/decision
Peter Drysdale	Founder/ visionary	Dignified, resilient housing at ultra-low cost	Defined model town concept; cyclone-resilient, modular design; sweat-equity build approach
Rotary Club of Lautoka	Institutional catalyst	Practical impact; credible governance	Seed funding, volunteer mobilisation, donor networks, early project stewardship
People's Community Network (PCN)	Resident governance body	Community voice, continuity	Social contracts, resident committees, dispute resolution, day-to-day operations
Fiji Government (local and central)	Regulator/ enabler	Public welfare, disaster risk reduction	Planning approvals, utility connections, targeted subsidies, policy recognition
International Rotary and Philanthropic Donors	Capital support	Scalable, high-leverage social return	Grants for construction phases, equipment, community facilities
NGOs (e.g., Habitat for Humanity)	Technical partner	Low-cost housing expertise	Community mobilisation methods, training, construction protocols
Residents (households)	Co-producers/ stewards	Secure tenure, stability, mobility	Sweat equity, maintenance levies, compliance with social contracts
Schools/Health and Social Services	Programmatic partners	Education and wellbeing outcomes	On-site classes, youth programs, health outreach, skills training
Disaster and Health Agencies (e.g., WHO links)	Resilience validators	DRR and public health gains	Post-disaster support, evidence for resilience and sanitation impacts
Local Suppliers/ Trades	Delivery ecosystem	Stable demand; skills transfer	Materials, training apprenticeships, ongoing maintenance capacity

3.3 Institutions of Resilience: Governance and Finance at Koroipita

Housing design alone cannot account for Koroipita's success. The institutions that support the initiative are what make it sustainable: an organizational model that combines resident participation, local leadership, and outside partnerships [1]; financial strategies that mobilize modest resources without succumbing to speculation; and governance mechanisms that inspire accountability in day-to-day operations. In a setting where many pilot programs end as donor interest wanes, these approaches collectively create the resilience architecture that has allowed Koroipita to survive for 20 years [9].

By design, the organizational model is hybrid. Since its founding, Koroipita has been more than a private experiment thanks to the institutional anchor supplied by the Rotary Club of Lautoka. Rotary's participation increased the project's respectability by connecting it to a worldwide network of contributors and volunteers. But as time went on, the PCN was used to localize governance. Additionally, PCN emerged as the official representative organization for the residents, serving as a forum for advocacy as well as a governing body. In addition to Rotary and PCN, the Fijian government has contributed in a limited way by offering regulatory control, infrastructure links, and certain subsidies [12]. Funding and technical assistance have been sporadically supplied by NGOs and international development organizations like Habitat for Humanity. The result is a partnership approach in which citizens' leadership is ingrained through PCN and committee structures, rather than any one body controlling the relationship.

The lease-based land arrangement in Koroipita is one of its unique features. Residents lease homes through long-term agreements with the governing trust instead of offering freehold titles, which run the danger of speculative resale and the deterioration of affordability [16]. This system maintains the project's overall culture while guaranteeing tenure, providing stability to families. Additionally, lease-based tenure guarantees that houses stay a part of the community's stock rather than being assets to be sold separately. By finding this equilibrium, Koroipita keeps the model's focus on need rather than profit and steers clear of the dangers of both commodification of affordable housing and unsafe squatting [17].

In terms of funding, Koroipita relies on a combination of resident donations, government subsidies, donor grants, and foreign aid. The capital costs of construction, especially in the early phases, have been covered by donor funds, frequently from Rotary chapters overseas [6]. The Fijian government has offered small subsidies for utilities like power and water connections. International relief organizations have occasionally made contributions, particularly following significant calamities when the project was hailed as a resilience model [18]. Affordable rental or lease fees, which are purposefully set below market rates yet adequate to pay upkeep and community initiatives, are one way that residents contribute. Although it is still susceptible to changes in donor support, this multi-stream funding strategy guarantees that the settlement is not entirely dependent on a single source of income.

Another layer of sustainability is added by cross-subsidization. Koroipita has established small-scale revenue-generating ventures that support the community's budget. For instance, recycling programs cut waste while making a small profit [1, 18]. In addition to improving livelihoods, training programs that help locals develop their skills in building, maintenance, and small-scale entrepreneurship also contribute to the town's upkeep [18]. These systems may not produce significant surpluses, but they lessen dependency on outside funding sources and promote a culture of independence.

A key component of daily management is participatory governance. Resident committees are in charge of managing upkeep, settling conflicts, and making sure that community rules are followed. Residents sign social contracts that include duties including keeping up gardens, paying rent, and taking part in group labour days. These systems establish a framework of reciprocal responsibility, whereby households are expected to contribute to the community's overall well-being in addition to enjoying safe, resilient housing. In addition to strengthening social cohesiveness and integrating stewardship into the project's DNA, the participatory method increases the potential for self-governance.

However, there are hazards associated with the financial and governance structure. Dependency on donors is still an issue. Despite the remarkable consistency of Rotary's global networks, the organization's reliance on charitable funding raises concerns about its long-term viability in the event that donor interest declines. The project's initial success was largely attributed to Peter Drysdale's vision and charisma, which raises concerns regarding continuity in his absence. The situation is made more complex by Fiji's political changes. Economic downturns or shifts in governmental objectives may result in less subsidies and weakened support for related programs. Lastly, a careful balance must be struck between the need to professionalize administration and maintain grassroots involvement. Insufficient institutionalization could expose the project to poor management, while excessive institutionalization could drive out people.

Koroipita's financial and governance structure is remarkably resilient in spite of these threats. The project has escaped the collapse that frequently befalls donor-driven schemes by integrating participatory governance, combining lease-based security with common ownership principles, and preserving a variety of funding sources. Its institutions embody the fundamental lesson of resourceful urbanism, which is that rules, relationships, and duties are just as important to resilience as walls and roofs. Despite their tiny size, Koroipita's finance and governance systems are an invention of worldwide significance because they demonstrate how communities with limited resources may create long-lasting futures.

3.4 Building Beyond Shelter: Implementation, Design, and Everyday Life in Koroipita

Deliberate phasing enabled the community to develop sustainably while incorporating lessons learned from each step into the subsequent one, which is how Koroipita was implemented. About 100 cyclone-resistant homes and basic infrastructure were supplied as part of Stage 1, which was started in 2003. These initial residences, which were constructed using a combination of expert advice and resident sweat equity, demonstrated that affordable, long-lasting housing was feasible in Fiji's resource-constrained environment. Stage 2 involved the construction of an additional 130 dwellings and enhanced community facilities after inhabitants had settled in, and governance mechanisms had stabilized. This gradual approach prevented overstretching, made sure that utilities like waste management, water, and sanitation kept up with population expansion, and gave locals the chance to improve their self-governance and upkeep techniques.

The concept of resourceful urbanism is reflected in the architectural principles that underpin these stages. The simple, modular homes were built for durability and efficiency rather than for show. They were resistant to cyclonic storms and heavy rainfall because of their lightweight but sturdy construction, reinforced anchoring, and straightforward designs. Self-help was ingrained as a social practice that strengthened group ownership and as a cost-saving tactic when construction processes mobilized resident labour. Repetition of modular designs allowed for gradual modifications when resources allowed, while also lowering costs and simplifying maintenance.

Crucially, Koroipita was never intended to be just a place to live. Every stage included communal infrastructure to facilitate daily living. By providing accessible education, schools and classrooms helped to lower the dropout rates that had been a problem in informal settlements. Meetings, training sessions, and cultural events were held at community halls. While creating modest revenue sources, an organized trash management and recycling system addressed environmental issues. In addition to providing food, community gardens promoted the virtues of independence and responsibility. Residents' health and dignity were guaranteed by giving priority to sanitation systems, which are frequently disregarded in informal communities.

The physical design was supported by programmatic initiatives. Opportunities for inhabitants were increased via education programs, which ranged from adult literacy classes to early childhood care. Initiatives for youth participation directed energy toward sports, cultural events, and skill development, which decreased social exclusion and paved the way for job opportunities. Through these initiatives, Koroipita was converted from a housing project into a vibrant, dynamic community.

Finally, resident-led maintenance programs helped to ingrain stewardship into the settlement's structure. Families agreed to pay small lease payments, participate in workdays, and contribute to communal maintenance as part of social contracts. These methods fostered a sense of accountability and community while ensuring that the built environment was maintained [18]. Thus, Koroipita's implementation serves as an example of how resilient, socially programmed, and modest physical improvements can create resilient, inclusive, and prosperous urban communities under pressure.

3.4.1 Resilience in Practice: Cyclone Winston and Beyond

One of the strongest cyclones ever recorded in the Southern Hemisphere was Cyclone Winston, which made landfall in Fiji in February 2016. With persistent gusts of above 230 km/h and deluges of rain, it wreaked havoc throughout the nation, demolishing or seriously damaging tens of thousands of homes, destroying infrastructure, and uprooting entire communities. Particularly severely affected were informal communities at Lautoka and beyond, where residents were compelled to spend months in makeshift shelters after roofs were torn off and lightweight housing was carried away. Koroipita stood out as a useful counterexample of resiliency in action amidst this extensive destruction.

Under Winston's harsh conditions, the settlement's homes—which were built with sturdy lightweight materials, low-profile construction, and reinforced anchoring—performed extremely well. In contrast to the devastating losses in many other informal communities, Koroipita's 230 dwellings sustained very minor damage, primarily to non-structural components like gardens, fencing, and outbuildings. Families managed to stay safe in their homes during the storm, and no houses fell apart. The project's guiding principle—that even inexpensive housing may be constructed to survive cyclonic forces if design and construction are appropriately tailored to local risks—was supported by this result. The event also refuted the idea that resilience must be prohibitively expensive, demonstrating that safety at scale may be achieved through creative design.

The timeliness and efficacy of community rehabilitation were just as crucial as the houses' physical functionality. Koroipita's governance structures were crucial. After the hurricane, resident committees mobilized labour, equipment, and expertise to expeditiously organize cleanup. Families were used to group responsibility and action due to the settlement's social contracts and maintenance schedules, which made mobilization during emergencies second nature. In order to manage debris, waste disposal and recycling systems were modified, and community halls served as centers of coordination. This group's resiliency stood in stark contrast to nearby informal settlements, whose recovery was more dispersed, slower, and significantly reliant on outside assistance.

The lessons are not limited to Koroipita. First, the case highlights the importance of including resilience into standard house design as a fundamental principle rather than as an add-on. Instead of only reacting to calamity, Koroipita avoided it by foreseeing risks in its building techniques. Second, social organization and governance are just as crucial as technical fixes [19]. Even with robust mechanisms, recovery would have been much slower without the existing structures of accountability, maintenance, and engagement. Third, Koroipita demonstrates that international assistance and state-led mega-projects are not the only options for climate adaptation in the Pacific. For disadvantaged communities, community-driven settlements that are gradually constructed with resilience at its heart can offer long-lasting safety nets.

After Cyclone Winston, Koroipita became a shining example of what resourceful urbanization can accomplish [1]. Cyclone Winston was a test by fire, or more accurately, by wind and water. Its record demonstrates that locally anchored, cooperatively sustained communities that turn constraint into capability can ensure resilience in the Pacific, rather than importing costly solutions.

3.5 Beyond the Village: Impacts, Limits, and Lessons from Koroipita

One of the Pacific's best-documented examples of community-led housing and resilience is Koroipita, which was established twenty years ago. Its effects are evident in quantitative, qualitative, and environmental metrics, and its drawbacks teach practitioners and policymakers equally important lessons. A fair assessment highlights the project's quantifiable benefits as well as the systemic issues that influence its long-term course.

On the quantitative side, Koroipita has delivered tangible results. The settlement, which housed about 1000 people, had more than 230 cyclone-resistant dwellings spread across two phases by the middle of the 2010s [1]. These families now have stable housing and tenure security, whereas previously they would have lived in unstable squatter communities. Supported by on-site classrooms and after-school programs, school attendance rates among Koroipita children increased dramatically when compared to those in surrounding informal settlements, further demonstrating the transformation. By giving adults and children the skills they need for small-scale entrepreneurship, maintenance, and construction, training programs have also opened up new options. Both directly and indirectly, since inhabitants are better equipped to seek steady employment due to their improved living conditions, jobs are created in the community through maintenance, daycare, and trash management positions. The recycling programme alone processes tonnes of waste annually, demonstrating how circular practices can underpin livelihoods while protecting the environment.

Qualitative results provide these figures more nuance. Testimonies from residents and interviews regularly indicate increased contentment with living conditions, a feeling of security during cyclone seasons, and enhanced dignity related to access to sanitary facilities, clean water, and well-organized communal life. The sense of belonging that the social contracts and participatory governance structures provide is arguably the most important. Residents participate in maintenance, decision-making, and cultural events as active participants in a group endeavour rather than just as tenants. This empowerment stands in stark contrast to the disempowerment that occurs in informal settlements, when confidence is undermined by neglect and insecurity.

Environmental factors set Koroipita out even further. As demonstrated by Cyclone Winston, when damage was modest in comparison to nearby villages, the settlement's resilient home design has significantly decreased disaster losses. Sanitation systems have reduced the incidence of waterborne illnesses, while recycling and waste management programs have enhanced the health of the environment. Reliance on outside markets has decreased and food security has been promoted by community gardens. When taken as a whole, these actions show how environmental sustainability can be included into situations with the fewest resources.

However, without addressing criticism, no assessment would be thorough. Koroipita's continued reliance on donor assistance, especially through Rotary networks, raises questions about its financial viability in the event that foreign support wanes. Centralizing leadership is another concern [18]. Peter Drysdale's leadership was crucial to the project's vision and cohesion, and it is unclear how governance will change without him. Another issue is scalability [6]. Although Koroipita is hailed as a model town, its replication throughout Fiji or the Pacific region necessitates land availability, political will, and consistent funding—all of which are uncertain. Critics further point out that although Koroipita provides protection to its citizens, it ignores the fundamental causes of informality, such as scarce urban jobs, complicated land tenure, and rural poverty.

Despite these drawbacks, Koroipita has shown to be remarkably resilient. After 20 years of constant use, growth, and modification, the model appears to be more than just a pilot—it is an institution. Less direct replication and more adaption of its fundamental ideas—lease-based tenure to avoid speculation, sweat equity to establish ownership, progressive phasing to manage resources, and participatory governance to maintain community life—are necessary for its transferability to other Pacific environments. Although they would need to be carefully adjusted to the local land regimes and political economy in each location, these ideas may be used in Port Moresby, Honiara [20], or Apia.

Thus, Koroipita is a warning as well as a light. It demonstrates how inventive urbanization may provide resilient, dignified, and long-lasting housing even in the face of severe limitations. However, it also serves as a reminder that scaling such ideas demands institutional backing, political will, and structural change in addition to vision. Its greatest contribution as a case study might be found in the lessons it teaches those attempting to create resilient, inclusive communities around the Pacific, as much as in the lives it has directly changed.

References

1. Prasad R, Muna L, Vodo S, Shivani, Azam MR, Charan D, Reddy A, Dean M, Asen M, Yaya F, Tamari L (2025) Development of waste management model for small Island communities: a case study of Koroipita, Fiji. Environ Sci Pollut Res, 1–15

2. Vunicagi KL (2016) Social housing in Fiji is social housing applicable for the poor and low-income only or does it include other levels of income earners? Master's thesis, Universidade do Porto (Portugal)
3. Pearce T, Currenti R, Doran B, Sidle R, Ford J, Leon J (2020) "Even if it doesn't come, you should be prepared": natural hazard perception, remoteness, and implications for disaster risk reduction in rural Fiji. International Journal of Disaster Risk Reduction 48:101591
4. Johnston I (2016) Rebuilding sustainable communities after disasters: Remote Islands. Cambridge Scholars Publishing, Cambridge
5. Loga PSI (2024) The evolution of public administration and conflict in a post-conflict state: history's role in Fiji's political trajectory: a thesis presented in partial fulfilment of the requirements for the degree of Doctor of Philosophy in Management at Massey University, Albany, New Zealand
6. Anaru K (2022) Designing for social impact. Doctoral dissertation, Open Access Te Herenga Waka-Victoria University of Wellington
7. Kundra S, Prasad RR, Alam MA, Alam M, Nobobo-Baba U, Ali MF, Iqubal A (2022) Urbanization in the Fiji Islands during the 21st century: a tourism perspective. In: Advances in urbanism, smart cities, and sustainability. CRC Press, pp 123–136
8. Reddy C (2020) Indo-Fijian fishing communities: relationships with Taukei in coastal fisheries. Doctoral dissertation, Open Access Te Herenga Waka-Victoria University of Wellington
9. Veitata SN (2023) A study of the potential for post-disaster resilience in indigenous Fijian communities. Available from https://repository.kulib.kyoto-u.ac.jp/dspace/bitstream/2433/283885/1/dtikk00238.pdf
10. Thalia OP (2024) Understanding the risk factors and vulnerable populations in the spread of diarrhea and typhoid fever: socioeconomic influences, malnutrition and susceptibility among risk groups. IDOSR J Appl Sci 9(3):1–6
11. Maiono JR (2024) Fifty years of independence: indigenous perception on sustainable land development in Fiji. Doctoral dissertation, University of Otago
12. Wei S (2012, July) Improving livelihoods through floriculture out-growing scheme-the Koroipita Village in Fiji. In: IV international symposium on improving the performance of supply chains in the transitional economies 1006, pp 405–410
13. McMichael C, Powell T (2021) Planned relocation and health: a case study from Fiji. Int J Environ Res Public Health 18(8):4355
14. Phillips T, Keen M (2016) Sharing the city: urban growth and governance in Suva, Fiji. Available from https://pacificsecurity.net/wp-content/uploads/2021/09/Sharing-the-City-Urban-Growth-and-Governance-in-Suva-Fiji.pdf
15. Cheshmehzangi A (2025) Social infrastructure: the measurable Core of Urban Society in Advanced City. In: Advanced City: for the future of cities and opposing smart cities. Singapore: Springer Nature Singapore, pp 145–154
16. Kiddle GL (2011) Informal settlers, perceived security of tenure and housing consolidation: case studies for urban Fiji. Doctoral dissertation, Open Access Te Herenga Waka-Victoria University of Wellington
17. Rantissi T, Gitis V (2024) Understanding squatting: an interdisciplinary analysis for effective policy interventions. Habitat Int 145:103012
18. World Health Organization (2019) Asia-Pacific parliamentarian forum on Global Health: fifth meeting: climate change and health, Nadi, Fiji, 20–22 August 2019
19. Cheshmehzangi A (2025) Service-driven governance: operationalizing advanced Urban Services for equitable access and performance. In: Advanced City: for the future of cities and opposing smart cities. Springer Nature Singapore, Singapore, pp 293–303
20. Keen M, Barbara J, Carpenter J, Evans D, Foukona J (2017) Urban development in Honiara. Harnessing opportunities, Enbracing change, state, society and governance in Melanesia program, Australian National University, Canberra. Published by Australian National University (ANU), 130 pages document

Part II
Adaptive Reuse and Retrofitting

Chapter 4
From Tracks to Commons: The Goods Line and Adaptive Urbanism in Sydney, Australia

4.1 From Freight to Footfall: The Political Economy of the Goods Line

Sydney's industrial background and its quest for new urban lifestyles in the twenty-first century meet with the Goods Line project. It is crucial to place an abandoned freight corridor within the larger historical and political economy of Sydney's urban change to comprehend how it evolved into one of the city's most renowned public areas [1]. The Goods Line, which was accomplished in the midst of changing economic structures, redevelopment pressures, and demands for public amenities in the city's crowded heart, was more than just a design experiment; it was a reinterpretation of industrial redundancy as municipal opportunity.

The freight rail system in Sydney in the late nineteenth and early twentieth centuries is where the corridor got its start [2]. In the past, the line carried everything from coal to agricultural products, connecting the busy port of Darling Harbour with the larger metropolitan rail network [3]. It served as a crucial conduit for decades in industrial Sydney, symbolizing the city's position as Australia's commercial hub and physically connecting production and consumption [4]. However, containerization and the shifting of port operations started to diminish Darling Harbour's importance by the middle of the twentieth century. The route declined as freight handling moved to bigger facilities at Port Kembla and Botany Bay [2]. The once-essential infrastructure was no longer in use by the 1980s; its tracks were rusting, its land was blocked off, and its contribution to the metropolitan economy was essentially obsolete.

This redundancy coincided with broader redevelopment schemes around Darling Harbour and Central Station. State governments embarked on large-scale initiatives to revitalize Sydney's inner core and waterfront in the 1980s and 1990s, turning post-industrial land into commercial, cultural, and recreational sectors [5, 6]. While the nearby Pyrmont and Ultimo neighbourhoods experienced waves of adaptive

A. Cheshmehzangi, *Resourceful Urbanism*, Urban Sustainability,
https://doi.org/10.1007/978-981-95-4138-6_4

reuse, from warehouses transformed into offices and studios to the growth of the University of Technology Sydney (UTS), Darling Harbour itself was transformed into a convention and tourism center [7]. However, in the midst of this change, the abandoned rail route continued to be a neglected piece. It cut across neighborhoods but made minimal contributions to urban life, acting more as a physical barrier than a link. It was both problematic and full of possibilities because of its prominence as a piece of "urban leftover" [8].

By the early 2000s, Sydney's urban agenda was increasingly shaped by concerns about liveability, connectivity, and sustainability. Governments were under pressure to develop pedestrian-friendly spaces and increase green infrastructure due to the inner city's growing population density, traffic, and scarcity of public space [6, 9]. International models like New York's High Line showed how abandoned train tracks may be creatively transformed into linear parks and public areas. Even while the Goods Line was smaller and had a smaller scope, it had the same symbolic potential—the ability to transform waste into value and redundancy into resource. In addition to providing better connectivity between Central Station and Darling Harbour, the corridor's reuse gave policymakers, designers, and organizations like UTS the chance to position Sydney as a city that values creative adaptive reuse [10–12].

It was crucial to reframe the corridor as a design issue. Planners and designers rethought it as infrastructure that already had significant material and historical worth, rather than as abandoned land that needed to be cleared or developed for commercial purposes. Instead of being removed, the linear geometry, heritage buildings, and embedded rail tracks might be kept and reimagined [5, 6]. This strategy was in line with a larger change in urban thought that prioritized low-cost, high-impact interventions above capital-intensive redevelopment and acknowledged the embodied energy in existing infrastructure.

The idea was made possible politically by a convergence of interests. With very small investments, the Sydney Harbour Foreshore Authority, on behalf of the New South Wales Government, attempted to revitalize the region. The Powerhouse Museum, the ABC, and UTS, among other nearby institutions, backed the corridor's makeover because it would increase accessibility and improve the public space surrounding their campuses [13]. Professionals in urban design, under the direction of Aspect Studios and CHROFI, advocated for a modern yet heritage-respecting approach [14]. These actors worked together to transform what might have been written off as a liability into a cherished civic spine [15].

Thus, the Goods Line is more than just a freight track turned pedestrian walkway. It serves as an example of how redevelopment pressures, cultural aspirations, and changing economic systems can come together to rethink obsolete infrastructure. Its development is indicative of a larger rebalancing of Sydney's political economy, moving from freight to foot traffic and from industrial throughput to urban liveability [16]. In this way, the Goods Line serves as both a local amenity and a prism through which to examine how cities coping with post-industrial legacies are changing, demonstrating how institutional cooperation and creative design can transform yesterday's waste into tomorrow's commons.

4.1.1 Forgotten Tracks: The Rail Corridor Before the Goods Line

The site that is currently known as the Goods Line was practically invisible prior to its makeover in 2015. Despite its prominent location in Sydney's downtown, extending from Central Station to Darling Harbour, it served more as a scar across the city than as connective tissue [5]. The corridor's physical and social circumstances in its abandoned state contribute to the explanation of why adaptive reuse turned out to be the most inventive answer and why it became a design issue that needed to be reframed.

Physically, the corridor bore the marks of redundancy. As port operations moved elsewhere, the freight line that had been transporting cargo into Darling Harbour for nearly a century was shut down. The line itself was gated off and closed to the public, its steel rails corroded, and its timber sleepers decayed. Through crevices in the ballast, vegetation began to grow, but it was weeds, scrub, and a reminder of carelessness rather than the lushness of a purposefully planned environment. The corridor, with its poorly illuminated areas and lightly enforced bounds, turned into an unofficial landfill in certain locations. The site provided no safe route for pedestrians, and for the city, it was a strip of land that was neither useable public space nor active infrastructure [6]. Safety hazards, from uneven surfaces to exposed rail remnants, underscored its unsuitability for casual access.

The train corridor was not seen as a social benefit, but rather as a barrier. It limited permeability in a region otherwise teeming with educational and cultural organizations by separating the UTS campus from Darling Harbour and the Powerhouse Museum sector [6]. The corridor caused pedestrians to take longer, less direct routes, reinforcing fragmentation rather than encouraging flow. In contrast to the surrounding modern structures, its visual presence—fences, rusty steel, overgrown edges—projected an air of neglect. This sense of disconnection was exacerbated by the corridor's very location: a linear void with no municipal significance in an increasingly crowded and sought-after area of the city.

The irony of neglect was heightened by the surrounding circumstances. In an effort to establish itself as a preeminent international university, UTS was growing its campus and making investments in new structures and amenities for students. Both staff and visitors were drawn to the Powerhouse Museum and the ABC, which maintained a prominent presence. At the end of the corridor was Sydney's busiest transportation hub, Central Station [17]. However, the train passage remained unoccupied, shuttered, and unutilized amidst this assemblage of public flows and institutions. In addition to being a waste of property, it was also a missed chance to establish a connection in one of Sydney's busiest neighbourhoods.

The constraints of traditional redevelopment choices, such removing the passageway to make room for new commercial buildings, were soon apparent. The site's tight geometry, heritage overlays, and proximity to significant municipal institutions would have made demolition and reconstruction unaffordable [6]. More significantly, these methods would have produced comparatively little in return and destroyed the corridor's industrial past. The surrounding precincts needed a connecting, accessible public space, which office towers and retail complexes could not have provided. Reimagination, not replacement, was required: a design strategy that could preserve heritage character, conserve embodied resources, and provide high-impact urban amenities with a small financial outlay.

In this way, the train corridor's pre-intervention circumstances were contradictory. The location was, on the one hand, abandoned, dispersed, and socially isolated. On the contrary, it was in a unique position to undergo change because of its location, linear shape, and heritage remnants. Its narrowness, industrial leftovers, and neglect—the exact characteristics that made it problematic—became the components of adaptive reuse. The Goods Line would subsequently show how urban waste may be repurposed as municipal infrastructure, mending what neglect had torn apart by transforming abandonment into opportunity.

4.2 From Proposal to Reality: Partnerships and Design Leadership Behind the Goods Line

It was not by accident that Sydney's abandoned freight corridor was converted into the Goods Line. It was the result of a meticulously planned process that involved the collaboration of institutions, community champions, design experts, and public agencies to transform an industrial waste into civic infrastructure. The project's emergence demonstrates that adaptive reuse necessitates coalitions of actors who are prepared to envision alternatives and align interests across scales, in addition to design inventiveness.

As the land custodian, the Sydney Harbour Foreshore Authority (SHFA), a division of the New South Wales Government, played a crucial role [18]. In addition to managing properties, SHFA was tasked with promoting urban redevelopment in areas of critical cultural and economic significance. By the late 2000s, the organization realized that the abandoned freight line that connected Central Station and Darling Harbour was not just a wasted resource but also a lost chance to improve connection in one of Sydney's most vibrant neighbourhoods [6]. In order to obtain early cash commitments and legitimize the initiative inside the government, SHFA's cooperation was essential. The corridor might have remained a fenced-off remnant of industrial decline in the absence of this institutional support.

The design process became the primary means of defining what the Goods Line could be after the political intent was formed. As principal designer, Aspect Studios, a landscape architecture firm known for creative public realm projects, collaborated with CHROFI (previously Choi Ropiha Fighera), an architectural studio that specializes in avant-garde urban intervention techniques [15]. Collectively, these companies put up a design that avoided the need to imitate international symbols such as the High Line in New York [19]. Rather, they looked for a distinctively Sydney solution: a simple, practical, and flexible makeover that respected the corridor's industrial past while satisfying regional demands.

Three guiding concepts formed the basis of the design concept. First, as part of the site's identity, preserve and reinterpret heritage traces, like embedded rail tracks. Second, the hallway might be transformed without requiring extensive construction by overlaying low-cost, modular interventions like timber benches, concrete planks, and shading devices. Third, make the most of programmatic flexibility by establishing areas for community gatherings, outdoor learning, and play that reflect the precinct's mixed institutional character [6]. This method was in line with the principles of resourceful urbanism, which include making the most of what already exists, making small but calculated investments, and planning for multiple uses rather than just one.

The project's justification and support were heavily reliant on local institutions. With its expanding campus and increasing student body, the UTS firmly supported the Goods Line as a secure, convenient pedestrian route from Central Station to its buildings. The corridor's ability to draw personnel and visitors through their precincts was also appreciated by the Powerhouse Museum and the Australian Broadcasting Corporation (ABC). In addition to providing convenience, the Goods Line was expected to improve the public space surrounding these schools' campuses, increasing their accessibility and visibility. Their backing made it possible to gain cross-sectoral legitimacy, guaranteeing that the project was presented as a civic investment rather than just a landscaping effort.

Professional and community advocacy also had a significant impact. The concept of linear parks had gained international attention due to the High Line's popularity in New York, and Sydney's design community was no exception [20]. The city's fast densification, according to urban planners, designers, and locals, necessitated the creation of new public spaces, especially ones that could connect disparate areas [14]. The Goods Line was purposefully scaled down and modified to fit Sydney's environment, even if the High Line served as a clear inspiration. The Goods Line sought to provide a practical, connecting spine for workers, commuters, and students, while the High Line developed into a tourist destination and a driving force for the development of the neighbourhood. This distinction was emphasized in the project's rhetoric, which emphasized that Sydney needed its own adaptive reuse model rather than a direct import.

These goals were strengthened by the competition and consultation process. In order to ensure that the design was co-produced rather than forced, Aspect Studios and CHROFI held consultations with stakeholders regarding programming, heritage preservation, and accessibility [6]. Community input focused on safety, illumination, and kid-friendly play areas, while workshops with institutional partners established the necessity of outdoor classrooms and learning facilities. The result was a design that was sensitive to the varied users of the corridor while also combining technological expertise.

The Goods Line had developed into a project with wide ownership by the time building started. Institutions like UTS, ABC, and the Powerhouse Museum served as anchor stakeholders; design companies converted vision into feasible form; the

NSW Government offered institutional legitimacy and finance; and community advocates gave moral and political impetus. The Goods Line was able to avoid the fate of several adaptive reuse initiatives that had stalled thanks to this coalition, proving that a small but impactful intervention may be realized through careful actor and interest alignment.

Thus, the rise of the Goods Line illustrates a more general reality of resourceful urbanism: creative design and the institutional frameworks necessary to facilitate it are both necessary for the repurposing of urban waste. Finding promise in an abandoned corridor is not enough; success depends on establishing the alliances that enable reuse, striking a balance between vision and pragmatism, and modifying international models to fit local conditions (See Table 4.1).

Table 4.1 Coalition to commons: who did what (and why)

Actor/Institution	Role in project	Core incentive	Key contribution/decision
Sydney Harbour Foreshore Authority (SHFA)	Land custodian; state lead	Revitalize strategic precincts with modest capital	Secured early funding, legitimized project, coordinated approvals
City of Sydney	Technical partner; public realm standards	Safe, maintainable, high-performing public space	Guidance on lighting, access, maintenance; integration with city networks
Aspect Studios	Lead landscape design	Demonstrate low-cost, high-impact reuse	Modular surface strategy; heritage rails embedded; programmatic layering
CHROFI	Architecture/urban design	Site-specific adaptive reuse (not a High Line clone)	Spatial sequencing; furniture systems; outdoor classrooms and play
UTS	Anchor stakeholder/user	Safe, direct link to campus; outdoor learning	Co-programming of study/teaching zones; continuous activation
Powerhouse Museum	Cultural precinct partner	Better access; audience growth	Event interfaces; wayfinding alignment with museum entries
ABC	Precinct institution	Staff/public permeability; civic frontage	Frontage improvements; passive surveillance and daily activity
Heritage NSW/ advisors	Heritage oversight	Retain industrial character; public safety	Keep/express rails and remnants; reconcile conservation with use
Community and Users	Everyday custodians	Shade, safety, play, informal use	Feedback on lighting, play, sightlines; shaped comfort and inclusion

4.3 Steering the Line: Governance and Finance Behind Sydney's Linear Park

The Goods Line is frequently praised for its sophisticated design and effective activation, but both governance and funding mechanisms were essential to its realization as it progressed from concept to reality [6]. Obtaining adequate institutional support and finance for a project that created public benefit but provided little immediate financial return was the problem, as it is with many adaptive reuse projects [5]. A pragmatic balancing act is reflected in the governance and financial framework that supported the Goods Line: public sector leadership, cautious integration into larger redevelopment objectives, and an understanding that this kind of municipal infrastructure is an investment rather than a source of revenue.

Leadership in the public sector was decisive. As the project manager and land custodian, the SHFA was established by the New South Wales Government. Enhancing public access to precincts of critical cultural and economic significance was part of SHFA's mandate, and the Goods Line fit in well with this goal. The City of Sydney contributed technical know-how on lighting, maintenance requirements, and public realm design. The project has the political clout to proceed and the local expertise to handle real-world implementation challenges thanks to this dual involvement—state-level authority and municipal support. Like so many underutilized infrastructure remnants, the corridor might have remained stuck in bureaucratic lethargy in the absence of its alignment.

The financing model was typical of urban design initiatives that had limited possibilities for direct money but strong symbolic significance. The Goods Line's design, construction, and initial landscaping were all covered by project-based budgets that were financed by state capital appropriations. Since the site's limited geometry and heritage status made commercial exploitation impractical, there was no need for continued land sales or private development contributions, in contrast to large-scale urban redevelopment programs. The project was instead viewed as a civic investment, similar to a cultural or educational building, and was supported more by place-making, connectedness, and public amenity than by financial gain. This framework was crucial because it made it possible to evaluate the Goods Line using community use, heritage interpretation, and accessibility indicators rather than profit margins.

The concept was further legitimized by being incorporated into larger redevelopment plans. Connecting Central Station, the focal point of the Central-to-Eveleigh urban regeneration corridor, with the Darling Harbour district, which has been the focus of numerous renewal waves since the 1980s, the route passes through a region that is undergoing substantial change [21]. SHFA made sure that the project was not viewed as a stand-alone experiment but rather as a component of a series of interventions meant to modernize Sydney's inner core by establishing the Goods Line as connecting infrastructure within these broader plans. Because the project could be positioned as enhancing tourism appeal and economic competitiveness in addition to its municipal duties [22], this integration also aided in obtaining funds and political support.

However, there were substantial governance issues. One of the main concerns was heritage conservation. The original rail tracks and related infrastructure in the corridor were valuable both materially and symbolically. In order to preserve these features as reminders of Sydney's industrial heritage while modifying them for safe, modern public usage, designers and authorities had to find a compromise [7, 23]. If the corridor had been preserved too much, it might have become a static relic; if it had been preserved too little, its unique character would have been lost. Although it needed careful negotiation between historical organizations, designers, and the state, the compromise that was ultimately chosen—integrating heritage rail tracks into the new surface treatments—proved to be an elegant solution.

Another difficulty was striking a balance between institutional demands and public use. Strong stakeholders including UTS, the ABC, and the Powerhouse Museum have a significant impact on the area that the Goods Line goes through. Each had justifiable concerns about visibility, programming, and access, but giving preference to one institution ran the risk of compromising the corridor's larger civic function. To ensure that the corridor functioned as a link rather than an extension of institutional branding, governance mechanisms had to uphold the concept that it was a public area first and foremost [6]. It wasn't always easy to strike this balance; there were complaints about how much the area favoured supporting nearby colleges. Nonetheless, the public nature of the broader governance paradigm was maintained.

The governance and financial narrative of The Goods Line emphasizes how crucial it is to present adaptive reuse as a form of public infrastructure that merits immediate funding. Sydney was able to complete a small but impactful project by fusing state-level leadership, local knowledge, incorporation into redevelopment plans, and cautious handling of stakeholder and heritage concerns [1, 2]. The necessity of maintaining authenticity while permitting new functions, the difficulty of defending investments without immediate financial returns, and the necessity of maintaining public value at the center of decision-making are just a few of the opportunities and conflicts that come with regulating adaptive reuse.

4.4 Implementation and Everyday Outcomes of the Goods Line

Through a series of well-considered architectural interventions, the Goods Line was transformed from a dormant freight corridor into a renowned urban spine without the need for costly reconstruction or extensive demolition. The project shows how even little physical components can have a disproportionately large social and cultural influence when added to already-existing infrastructure. Its execution, urban planning techniques, and programmatic results are prime examples of resourceful urbanism in action: repurposing heritage elements as the foundation of a multipurpose public area and creating more out of less.

At the heart of the project were lightweight physical interventions. The hallway was treated as a linear platform by Aspect Studios and CHROFI, who covered it with modular concrete planks that served as areas for play, seating, and casual gatherings in addition to circulation routes. By softening the sharper edges with timber components, benches and integrated furniture were created that encouraged people to stop rather than just move through. The hallway was designed with play places, such as open spaces and climbing frames, where kids could engage with the surroundings [6]. Strategically positioned lightweight shading devices provided relief from Sydney's heat without overpowering the corridor's constrained layout. Crucially, the project was grounded in its industrial past without resorting to nostalgia thanks to the preservation and embedding of heritage rail tracks into the surface. These subtle yet expressive design choices provided identity and utility without going over budget.

The main driver for the initiative was connectivity, which soon shown its worth. The Goods Line offered a direct, excellent connection for bicyclists and pedestrians between Darling Harbour, one of the city's most popular entertainment and cultural districts, and Central Station, Sydney's main transportation hub. By connecting organizations like the Powerhouse Museum, the ABC, and UTS, the corridor transformed what had been a barrier into a common core [2]. Increases in foot traffic were documented, indicating the corridor's function as a regular commute route for workers, students, and visitors. By putting mobility and accessibility first, the Goods Line increased public space while delivering efficient transportation.

Beyond connectivity, the corridor's value was expanded through programmatic outcomes. Because of the outdoor classrooms and learning areas included in the design, UTS is able to extend its educational operations into public areas [24]. The corridor's linear shape and institutional proximity were utilized for cultural programming, ranging from temporary performances to art installations. The area was brought to life by events and unofficial recreation, making it serve as more than just a road; it was a destination unto itself. Office workers took refuge on the benches, families played in the play areas, and tourists used the corridor as a gateway to Sydney's cultural center. The project's vitality depended on the stacking of programmed functions.

The Goods Line's low-cost, high-impact approach was arguably its most inventive feature. Adaptive reuse was adopted by designers in place of demolition and reconstruction, saving embodied energy and lowering building costs [6]. Due of the corridor's short 500-meter length, even little changes had a significant impact and changed how people perceived the entire path. The Goods Line showed how focused, incremental design could provide connectivity, heritage interpretation, and civic amenity at a fraction of the cost of large-scale urban park initiatives that required enormous capital outlays.

The Goods Line's implementation confirmed a notion that is frequently disregarded in urban redevelopment: that using what already exists can create public benefit. A space that is adaptable, sturdy, and unique was produced by the minimalist material palette, multipurpose programmatic design, and focus on connectivity. It is a notable illustration of how adaptive reuse can turn abandoned infrastructure into a stage for urban activity, even though it is not massive.

4.4.1 Walking the Line: Everyday Use and Public Reception

The Goods Line has grown to be an essential component of Sydney's public space since it opened in 2015, not because of its size but rather because of how it has been integrated into everyday routines [6]. Its viability as a 500-meter link depends on commuters, students, employees, and tourists constantly using it to pass through, stop in, and engage with what was formerly an abandoned freight route. The best proof that the project has fulfilled its promise of adaptive reuse comes from daily patterns of activity.

The most concrete measure of effectiveness is the documented improvements in cycling and pedestrian movements. By connecting Central Station and Darling Harbour directly, the corridor shortens travel times and promotes active transportation. According to surveys and observational studies, there is a sizable daily foot traffic, with University of Technology Sydney students making up the main user base. While office workers in surrounding buildings use the Goods Line for informal meetings and lunch breaks, it has become a vital route for them to get to campus. Its dual purpose as a commuter spine and leisure amenity is highlighted by the fact that tourists who are visiting the Powerhouse Museum or the waterfront precinct also incorporate it into their city experience.

User feedback demonstrates its adaptability. Students appreciate how the design incorporates casual study areas and outdoor classrooms. Play areas that give kids secure, enclosed spaces are valued by families. Shaded seating is a favourite spot for workers to rest and rejuvenate throughout the day. The integrated rail tracks and heritage allusions, which give the area personality and authenticity, are often noted by visitors. This variety of applications shows how the project's little design changes have resulted in disproportionately rich results.

The professional community has also embraced the Goods Line. The project has received numerous awards, including the Australian Urban Design Awards and recognition from the Australian Institute of Landscape Architects (AILA). These accolades reflect the project's significance as a benchmark for adaptive reuse in Australia, applauded for its ability to balance heritage conservation, contemporary design, and functional performance.

However, there have been criticisms of the public's response. When contrasted with the more verdant settings of Sydney's other parks, some observers draw attention to the corridor's sparse vegetation. There has been discussion on whether the Goods Line effectively handles urban heat and ecological functions because it is more hardscaped than green. Others doubt its scope, asking whether it is more symbolic than revolutionary at a mere half-kilometer. Some view the initiative as a potent symbol of Sydney's capacity to reuse urban waste, while others believe it is too brief and understated to change the paradigm of public space in the city.

Nonetheless, the Goods Line is instructive in part because of these criticisms. Its welcome demonstrates that adaptive reuse initiatives can have significant symbolic significance and promote everyday usage, even at tiny scales [6]. The hallmark of a project that has truly penetrated the city's cultural and civic discourse is that it is both praised and questioned.

4.5 Beyond the Corridor: Impacts, Limits, and Lessons from the Goods Line

The Goods Line provides a strong example of how even small adaptive reuse may transform everyday urban life, but its impact must be evaluated based on evidence-based results as well as aesthetic appeal and prizes [6]. The quantitative, qualitative, and environmental effects of the project demonstrate how a brief corridor has produced a wide range of public benefits. However, criticisms of its scope and ecological complexity draw attention to the limitations of these kinds of initiatives. When taken as a whole, these factors place the Goods Line in the context of larger discussions concerning resourceful urbanism and the future of urban adaptive reuse.

The Goods Line has clearly impacted the flow of people walking and bicycling, according to quantitative data. Following its inauguration, the corridor saw a significant increase in foot traffic, according to studies conducted by the NSW Government and outside observers. Some trips were switched from vehicle to active modes, and travel durations were shortened by the direct connection between Central Station and Darling Harbour. Although its 500-meter length restricts its potential to influence the shift in the city's modal share, the Goods Line greatly enhanced walkability within its catchment. It fared better than traditional park provision in terms of cost-benefit analysis. It was an effective investment in public infrastructure because to its high usage rates and connectivity advantages, which were achieved at a relatively cheap capital outlay when compared to newly constructed open spaces.

Qualitative effects are just as important. By converting a forgotten industrial scar into a cherished urban treasure, The Goods Line has created a feeling of place. In addition to preserving memories, its integrated train tracks and industrial allusions provide the district a new cultural character. It has developed into a symbolic front yard and connector for organizations like UTS, ABC, and the Powerhouse Museum, fortifying connections between hitherto disparate locations [11]. Users frequently express emotions of enjoyment, safety, and accessibility, indicating that the corridor has changed how people view the area. In this sense, the Goods Line serves as a cultural spine in addition to a route.

Even though they are less significant, environmental results are nonetheless significant. The project saved embodied carbon that would have been lost during demolition and reconstruction by reusing the existing rail line and keeping heritage materials. By providing shade and lowering heat in certain areas of the corridor, incremental greening—planter boxes, shade trees, and permeable surfaces—introduced minor microclimate benefits. However, detractors frequently emphasize that the environmental impact is still minimal when compared to more ecologically ambitious urban parks.

These criticisms are necessary for a fair evaluation. Unquestionably short, the Goods Line cannot completely replace larger parks or ecological corridors due to its compactness. Some have expressed disappointment with the relative lack of vegetation, despite the hardscaped character's deliberate emphasis on industrial past. Others caution against over-programming, pointing out that putting study, play, and cultural areas on such a short section could lead to rivalry for available space. However, rather than being interpreted as failures, these restrictions serve as a reminder of the project's precise goals and scope. It was intended to be a connecting thread sewn into an existing dense fabric rather than Sydney's Central Park.

Transferability and longevity complete the importance of the Goods Line. The corridor's continued success as a regularly used municipal spine nearly ten years after it was completed shows how resilient lightweight adaptive reuse can be. The lesson is not one of duplication but rather one of principle for other towns in Australia and around the world. If the public benefit is prioritized, stakeholder coalitions are strong, and the design is modest, adaptive reuse can reveal latent value in underutilized infrastructure. Whether in global environments like Toronto or São Paulo, or in Melbourne's railyards or Brisbane's abandoned industrial corridors, the Goods Line provides an example of how "urban leftovers" can be transformed into civic commons that foster connection.

Finally, the Goods Line should be viewed more as a proof of concept than a final result, showing that even the tiniest pieces of infrastructure can be turned into significant urban assets. Its quantified results demonstrate that, at a fraction of the expense of new creation, resourcefulness—making do with what currently exists—can create everyday usage, identity, and connectivity. Its criticisms serve as a reminder that while adaptive reuse can effectively and elegantly manage many urban issues, it cannot resolve all of them. The Goods Line epitomizes the spirit of resourceful urbanization in its harmony of accomplishment and constraint: modest in means, generous in impact.

References

1. NSW Government (n.d.) The Goods Line. Available from https://www.nsw.gov.au/visiting-and-exploring-nsw/locations-and-attractions/goods-line
2. Hansen SI (2004) Transportation and urban development: the past and future of Sydney. Doctoral dissertation, University of Wollongong
3. Symes, C. (2007). On the right track: railways and schools in late nineteenth century Sydney. In: Spatial theories of education. Routledge, New York, pp 175–194
4. Bunker R, Freestone R, Randolph B (2017) Sydney: growth, globalization and governance. In: Planning metropolitan Australia. Routledge, New York, pp 76–100

5. Vivid Sydney (n.d.) *The Goods Line and Inner City*. Available from https://www.vividsydney.com/locations/the-goods-line
6. NSW Government (n.d.) The goods line. Ultimo: Case study. Available from https://www.planning.nsw.gov.au/government-architect-nsw/case-studies/the-goods-line-ultimo
7. Kaya E (2020) Transformation of Sydney's industrial historic waterfront. Springer, Singapore
8. Cheshmehzangi A (2025) Australia's urbanity: architecture, place, and the pulse of urban life. Springer, Singapore
9. Searle G, Pritchard B (2008) Beyond planning: Sydney's knowledge sector development. In Knowledge-based urban development: planning and applications in the information era. IGI Global, pp 184–202
10. Alic D (1992) Old transport buildings and sites: re-use for transport purposes or adaptive re-use for other than transport purposes. Doctoral dissertation, UNSW Sydney
11. Bennett J, Beudel S (2014) Curating Sydney: imagining the city's future. UNSW Press, Sydney
12. Kean G (2015) Maximising the potential of existing urban infrastructure: can infrastructure reuse provide successful public spaces? Doctoral dissertation, University of Otago
13. Freestone R, Pullan N, Saniga A (2021) The making of a city campus. Geogr Res 59(1):29–45
14. Heritage Council of NSW (2019) Design Guide for Heritage: Implementing the Better Placed Policy for Heritage buildings, sites and precincts, Issue no. 2. Available from https://www.environment.nsw.gov.au/sites/default/files/better-placed-heritage-design-guide.pdf
15. Cheshmehzangi A (2025) Advanced City: for the future of cities and opposing smart cities. Springer, Singapore
16. Bunker R, Troy L (2015) The changing political economy of the compact city and higher density urban renewal in Sydney. Planning in a Market Economy–Working Paper 1
17. Farrelly E (2021) Killing Sydney: the fight for a City's soul. Picador Australia, Sydney
18. Hoskins I (2009) Sydney harbour: a history. UNSW Press, Sydney
19. Huebner J (2019, February) High line in New York–an unusual park in the Centre of Manhattan. In: IOP conference series: materials science and engineering, vol 471, 9, p 092085. IOP Publishing
20. Loughran K (2016) Imbricated spaces: the high line, urban parks, and the cultural meaning of city and nature. Sociol Theory 34(4):311–334
21. Rauscher R (2021) Inner City Renewal and Value Capture Planning–Greater Sydney's Central to Eveleigh Renewal Area (CERA). In: Renewing cities with value capture planning: model for achieving equitable housing, public and open spaces, and sustainable transport. Springer International Publishing, Cham, pp 55–88
22. Cheshmehzangi A (2020) Reflecting on urban identities. In: Identity of cities and City of identities. Springer Singapore, Singapore, pp 133–178
23. Punter J (2004) From the ill-mannered to the iconic: design regulation in central Sydney 1947–2002. Town Plan Rev 75(4):405–445
24. Gusheh M, Firth V, Netherton C, Pettigrew C (2019) The creation of the UTS social impact framework: a collaborative approach for transformational change. Gateways Int J Community Res Engagem 12(2):1–22

Chapter 5
From Grain to Gallery: The Zeitz MOCAA and the Cultural Rebirth of Cape Town's Waterfront

5.1 From Empire's Granary to Africa's Gallery: Context and Political Economy of the Zeitz MOCAA

In Cape Town's cultural and urban history, the Zeitz Museum of Contemporary Art Africa (MOCAA) holds a unique place. Without first retracing the development of the city's waterfront, the industrial and political significance of the grain silo it occupies, and the larger dynamics of post-apartheid South Africa's quest for cultural representation on a continental scale, it is impossible to comprehend its emergence [1]. This background demonstrates how one of Africa's most significant cultural initiatives was launched from a piece of abandoned industrial infrastructure.

At the Victoria & Alfred (V&A) ports, a thriving center of maritime commerce during British colonial authority, the grain silo complex was built in 1921 [2]. The silo served as a representation of South Africa's agricultural prosperity and its incorporation into international commodities markets for a large portion of the twentieth century. For many years, it dominated Cape Town's skyline at 57 meters, making it the highest structure in sub-Saharan Africa. Millions of tonnes of grain meant for export were kept in its 42 concrete cylinders, connecting the country's rural hinterlands to global markets [3]. The silo represented the economic goals of a nation whose wealth was based on agricultural exports and extractive industries, making it more than just an industrial device. However, it was intertwined with systems of inequality, favouring white commercial farmers while marginalizing Black rural inhabitants, much like a large portion of South Africa's infrastructure in the twentieth century [4].

The silo's fortunes had deteriorated by the end of the twentieth century. The structure became outdated due to changes in shipping technologies, containerization, and the transfer of bulk handling to larger deepwater ports. The silo, an imposing reminder of a bygone industrial era, was gradually abandoned as grain was no

A. Cheshmehzangi, *Resourceful Urbanism*, Urban Sustainability,
https://doi.org/10.1007/978-981-95-4138-6_5

longer transported via the V&A docks. Its shuttered cylinders and rusted machinery served as a metaphor for the global demise of operational waterfronts.

The surrounding waterfront was also going through significant alteration at the same period. Redevelopment of the V&A Waterfront, one of South Africa's premier urban regeneration projects, started in the late 1980s and picked up speed following the democratic transition in 1994 [4]. The region was transformed into a mixed-use precinct that combines office space, retail, leisure, and tourism after previously being a zone of warehouses, shipyards, and dock labour. In post-apartheid South Africa, this redevelopment held particular symbolic weight, although it also followed worldwide trends in the adaptive reuse of industrial waterfronts, from Baltimore to Barcelona. The V&A Waterfront was intended to be Cape Town's international entrance, giving the city a sense of openness and modernity. Millions of local and foreign tourists visited it every year by the 2000s, making it Africa's most popular travel destination. The grain silo, however, remained intact despite this success because it was too intricate to be readily used and too famous to be destroyed.

In terms of culture, the lack of a significant organization devoted to modern African art has grown more noticeable [5, 6]. Although there were thriving art scenes in Cape Town and Johannesburg, exhibitions were frequently dispersed among little galleries, private collections, and temporary displays. With a population of more than a billion, Africa lacked a specialized museum to display its modern innovation on the same scale as New York's MoMA or Europe's Tate Modern. Deeper disparities in global cultural infrastructures were reflected in this absence, since African artists frequently achieved international acclaim before receiving domestic acclaim. It was obvious that an organization was required to bring African modern art together, give it legitimacy, and present it to the world.

The Zeitz MOCAA emerged as a result of the convergence of several factors: a cultural void, an abandoned industrial landmark, and a shoreline that was repositioned for global consumption. The massive silo in the center of the V&A Waterfront presented both a problem and a chance [6]. Adaptive reuse was a cost-effective alternative to demolition that complied with sustainability requirements. A significant shift in values from grain to gallery, from resource export to artistic expression, was politically signalled by the transformation of a remnant of extractive commerce into a venue for cultural celebration. The change represented a continent regaining its industrial infrastructure for cultural sovereignty in a symbolic sense (see Table 5.1).

Therefore, the Zeitz MOCAA accomplished more than only close a gap in South Africa's cultural environment. It emerged from a specific political economy that was influenced by global cultural injustices, urban renewal, industrial decline, and the symbolic value of adaptive reuse. Its background serves as a reminder that museums are not neutral spaces but rather participants in historical and urban conflicts, having connotations that go well beyond their physical boundaries [7]. The transition from an empire's granary to Africa's gallery in the grain silo on Cape Town's waterfront captures the paradoxes and potential of creative urbanization in the Global South.

Table 5.1 contrasts the grain silo's industrial past with its cultural rebirth as Zeitz MOCAA, highlighting the symbolic shift from exporting raw resources to showcasing African creativity and sovereignty

Dimension	Grain Silo era (1921–1990s)	Gallery era (2017–Present)
Economic role	Agricultural export hub; stored millions of tonnes of grain; integrated South Africa into global commodities markets	Anchor for cultural tourism; generates revenue through exhibitions, events, retail, and precinct synergy
Political symbol	Colonial and apartheid industrial infrastructure privileging white commercial farmers	Pan-African cultural institution reclaiming space for African art and sovereignty
Urban identity	Industrial skyline marker; functional infrastructure disconnected from civic life	Iconic landmark within V&A Waterfront's mixed-use redevelopment; civic and cultural anchor
Global connection	Exporting raw resources to foreign markets; dependency on extractive industries	Showcasing African creativity to global audiences; recognition of African art on its own soil
Social meaning	Symbol of inequality and exclusion; inaccessible to most of the population	Space of visibility, education, and cultural pride—yet still facing debates on accessibility and elitism

5.2 Coalitions of Culture: The Making of the Zeitz MOCAA

Instead of being the result of the grain silo's collapse, the Zeitz MOCAA was the result of a carefully curated coalition of actors, including philanthropists, developers, architects, and cultural advocates, who came together with the goal of turning an abandoned industrial landmark into a cultural hub for the continent [8]. The narrative demonstrates how adaptive reuse on this scale necessitates more than just creative design; it also depends on coordinating institutional interests, private resources, and more general symbolic goals.

Jochen Zeitz, a German-born businessman best known for his time as Puma's CEO, was at the center of this alliance. Zeitz had already made a name for himself as a patron and collector of contemporary African art, amassing a sizeable private collection and establishing charitable endeavours through the Zeitz Foundation. His dedication was not just materialistic; rather, it was presented as a component of a larger initiative to promote African creativity, cultural expression, and sustainability [9]. Zeitz saw the absence of a significant museum devoted to modern African art as a chance as well as a cultural injustice. Together with the funding for its establishment, his choice to offer his collection as the foundation of a new organization served as the catalyst for the project's development.

However, Zeitz's concept needed an institutional partner and a proper home. This is where one of Africa's most popular tourist spots, the V&A Waterfront, came into play. The administration of V&A realized by the middle of the 2010s that although the precinct was profitable, it ran the risk of becoming overrun with retail and recreational establishments. The development required an international cultural anchor in order to maintain its reputation and broaden its appeal [10]. Located prominently within the precinct, the abandoned grain silo provided the ideal symbolic location for the establishment of such an organization. In order to strengthen the V&A Waterfront's reputation as a civic and cultural hub in addition to a retail destination, the partnership with Zeitz pledged to strike a balance between business and culture.

To transform this idea into a feasible enterprise, the design commission was essential. The British designer Thomas Heatherwick led the Heatherwick Studio, which was tasked with transforming the silo's thick concrete cylinders into useful gallery areas [11]. The adaptive reuse language served as the foundation for Heatherwick's strategy, which aimed to create strikingly new spatial experiences while preserving the industrial feel [12, 13]. This idea was embodied in the center atrium, which was cut out of the silo's bins to create a cathedral-like interior. In order to strengthen the museum's international design heritage, technical assistance and involvement were sought from Herzog & de Meuron, the Swiss architects known for cultural initiatives like Tate Modern [14]. Beyond a utilitarian conversion, the project's design ambition elevated it to the status of a global architectural statement that could coexist with other iconic institutions.

Funding, governance, and curatorial vision negotiations were crucial behind the architectural drama. The museum needed a framework that would guarantee long-term financial sustainability while striking a balance between Zeitz's philanthropic contributions and institutional autonomy. After agreements were reached, Zeitz donated his collection and an endowment to fund the museum's operations, putting it under a not-for-profit trust. The building shell and continuous integration within the precinct were provided by the V&A Waterfront [4]. This arrangement served as an example of a hybrid governance approach, with public access and institutional frameworks providing civic legitimacy while private philanthropy served as the initiative's anchor.

The project's goals went beyond just giving one man's collection a place to call home. Establishing the first significant pan-African institution devoted to contemporary art from the continent and its diaspora was the proclaimed goal from the beginning. This was an important framing. It aimed to rectify the disparity that African art was frequently displayed in American or European institutions instead of on African territory. This contradiction was echoed in the discussions surrounding governance and curatorial strategy: how to make sure the museum served as a platform for African artists, curators, and audiences rather than just a monument to the idea of a European collector. The creation of this concept was a crucial step in securing the museum's validity, even though discussions over accessibility and inclusivity are still ongoing.

A philanthropist with a collection and a mission, a developer looking for a cultural anchor, architects who could turn an industrial relic into an icon, and governance structures that sought to balance private initiative with public responsibility were all involved in the creation of the Zeitz MOCAA [1]. The alliance that created the museum exemplifies the complexity of the Global South's cultural infrastructure: it is reliant on international funding and design know-how while attempting to express local identity and continental aspirations.

5.2.1 From Derelict Silo to Cultural Landmark

It was not inevitable that Cape Town's abandoned grain silo would become the Zeitz MOCAA. For many years, the building was seen as a barrier in the renovated V&A Waterfront—too big to overlook, yet too ungainly and expensive to modify [1]. Like many abandoned industrial sites across the world, early plans tended toward demolition or commercial rehabilitation. Some developers proposed demolishing the building to create space for mixed-use skyscrapers, hotels, or shopping centers. Others thought about partially demolishing the silo to preserve pieces as heritage landmarks and raise money for private investment. These choices ran the risk of destroying a piece of Cape Town's cultural legacy and one of the city's most recognizable industrial monuments [15], but they also promised financial success.

The realization that adaptive reuse could save environmental costs, preserve heritage, and create a distinctive cultural attraction all at once was a conceptual leap. Once a drawback, the silo's striking concrete geometry was reframed as its greatest strength. In order to create new galleries, Heatherwick Studio suggested hollowing out the structure's cylindrical bins while keeping the dense fabric intact [1]. They turned storage cylinders into a sculpture that recalls a cathedral and a cavern by carving a large central atrium out of the group of silos at the center of the project. By this action, the structure was reinterpreted as a raw material for architectural innovation rather than as a remnant to be replaced.

The technical difficulties were severe. Precision engineering was needed to cut into concrete tubes that were 33 meters high while maintaining structural integrity for additional loads, such climate-controlled galleries [1]. It took creativity to weave circulation routes through the silo's dense matrix, integrate glass elevators, and reinforce walls. The question of aesthetics was as urgent: how to honour the industrial spirit without overpowering guests with unadulterated utilitarianism. The answer was a design language that incorporated modern features like sculptural stairs and streamlined glazing while highlighting the silo's geometry, which included polished concrete, curved voids, and filtered light.

Although controversial, this design concept—keeping the industrial outside while creating new symbolic interiors—was eventually revolutionary. Proponents viewed it as a potent illustration of resourceful urbanism, preserving embodied energy, incorporating cultural significance into a location that was formerly used for export and extraction, and working with inherited structures rather than destroying them. The extent of the alteration, according to some, undermined authenticity and made the silo a stage prop rather than a relic that had been conserved. Whether adaptive reuse should preserve the past or rework it for modern relevance was at the center of heritage arguments.

Finally, the project's solution veered more toward reinvention than rigid conservation. The silo was transformed into a museum and a place of recollection by drastically altering its interior while maintaining the building's exterior silhouette and industrial traces. An industrial remnant transformed into Africa's largest museum of contemporary art; it became not only a landmark for Cape Town but also a symbol for the continent.

5.3 Governing Culture: Architecture and Finance of the Zeitz MOCAA

The Zeitz MOCAA's financial and institutional architecture is just as intricate as the structure itself [1]. Large-scale adaptive reuse initiatives rarely succeed based solely on design vision; they need strong governance structures and sustainable financial models that can ensure the institution's long-term viability. The foundation of Zeitz MOCAA was established through a hybrid framework: a public-private partnership that combined developer investment, donor finance, and operational revenue streams. This model was overlaid with discussions regarding African ownership and curatorial independence.

The V&A Waterfront, the developer and landlord in charge of a large portion of Cape Town's famous harbor district, is at the center of the governance paradigm. The V&A Waterfront served as the museum's physical location, owning the silo building and maintaining the public space around it. It had a strategic interest since it raised the precinct's status as more than just a shopping and tourism destination by integrating a cultural anchor. With the museum positioned as a cultural and civic addition to its business endeavours, the V&A made sure the project stayed linked to the waterfront's long-term revitalization by agreeing to place it in the silo.

The catalytic spark came from the Zeitz Foundation. The first cultural and financial credibility was provided by Jochen Zeitz's contribution of his private collection of modern African art along with an endowment to support operations [16]. With this gift, the silo was turned from an empty shell into a storehouse for a story about pan-African art. However, the foundation's involvement also brought with it the risk of becoming overly dependent on the finances and vision of a single patron; this conflict was resolved by establishing a non-profit trust to handle the museum's operations.

To strike a balance between philanthropy and operating revenue, funding sources have been purposefully diversified. With the help of sporadic state and foreign development funds, the museum's income is derived from ticket sales, membership programs, events, and retail activities [1]. Given the significant operating expenditures of the museum—from delivering large-scale shows to maintaining climate-controlled galleries—these many revenue streams are crucial. Cross-subsidization is involved: business gatherings and tourism ticket sales support educational initiatives and community service, which is consistent with a more comprehensive approach of cultural resourcefulness.

The area of governance has been contentious. With a board of members from the corporate, civil society, and art sectors, the museum trust was created to protect the organization from private takeover. This architecture has been put to the test by leadership changes, with high-profile resignations and discussions about curatorial independence garnering attention from all around the world. Given the museum's assertion that it serves the African continent as a whole, questions of representation—whose voices and artistic visions are amplified—have been particularly pertinent. It has consistently been difficult to guarantee that governance arrangements continue to be transparent, responsible, and inclusive.

It is evident that this architecture has inherent hazards. The museum is susceptible to changes in donor interests because of its reliance on private philanthropy. In a district that is primarily driven by tourism, striking a balance between cultural agendas and commercial imperatives is a constant negotiation. Above all, the issue of African ownership and representation is still crucial: how can a museum established with the help of international architectural knowledge and European charity genuinely ground itself in African cultural sovereignty?

However, Zeitz MOCAA's financial and governance structure exemplifies a practical kind of ingenious urbanism [17]. By combining developer integration, operational diversity, and philanthropic donations, the museum has produced a platform that is a combination of state-run and commercial operations. Although it is a delicate balance, it has allowed the silo to reimagine itself as Africa's premier museum of modern art.

5.4 Carving Culture: Implementation, Design and Programmatic Outcomes

Not only did Cape Town's grain silo need to be transformed into the Zeitz MOCAA, but it also needed a design strategy that could repurpose industrial infrastructure as civic architecture [1]. A new cultural icon for the continent can be created by combining architectural innovation, urban integration, and programmatic depth, as demonstrated by the silo's adaptive reuse.

The central atrium, an ambitious intervention that gave the building its worldwide architectural signature, is at the center of the design. The team at Heatherwick Studio cut and polished the concrete to expose the tubular geometry of the cylindrical bins in a sculptural cross-section, creating a huge void from the dense cluster. This central area transforms raw industrial mass into a cathedral-like volume of light and shadow, serving as a hub for circulation as well as an iconic architectural gesture. The building's dual identities as heritage and reinvention are reinforced by spiral staircases that wrap around its edges and glass elevators that rise through the hollowed emptiness. Subtraction, or the removal of material to create space, became the defining technique for the design.

More than 80 galleries are housed within the silo's skilfully modified fabric beyond the atrium. While bigger open levels above the silo foundation allowed for expansive exhibition halls, individual bins were transformed into smaller galleries perfect for focused pieces. Alongside galleries, education centers, conservation labs, and research facilities were added to ensure the museum could serve as a knowledge center in addition to a place for displays. Each intervention expanded upon the preexisting structural grid rather than completely replacing it, demonstrating a creative use of space in this programmatic layering.

A primary goal of implementation was integration with the larger V&A Waterfront precinct [2]. Because of its advantageous location between the working harbor and the commercial waterfront, the museum attracts foot traffic from shopping districts and connects people to civic and cultural events. The museum was positioned as a destination and a connector by incorporating new pedestrian linkages within the precinct [18]. Although the silo's façade remained a recognizable landmark on the skyline, its reimagining represented a more significant change: from the export of commodities to the creation of culture, from industrial circulation to public gathering.

This change was further supported by programmatic results. Zeitz MOCAA has hosted significant exhibitions by African and diasporic artists since its opening in 2017, giving them exposure that has been long denied in international cultural circles. In addition, the museum hosts a wide range of public events, such as artist residencies and educational seminars, establishing itself as a venue for both elite cultural exhibition and community involvement. Its educational programs emphasize art as a vehicle for identity construction and critical thinking, and they are aimed at young South Africans [1, 2]. In the meanwhile, mentorship programs and residencies help up-and-coming African artists by resolving systemic disparities in recognition and opportunity.

There were difficulties in implementation. Careful negotiating was necessary due to technical intricacy, financial constraints, and discussions about curatorial direction. However, the end result shows that adaptive reuse can accomplish more than just preserving materials; it can also enable new futures and build environments that reflect memories. The galleries, the public programs, and the carved atrium all show how an old grain silo was converted into a living cultural infrastructure.

The application of Zeitz MOCAA highlights the ideas of resourceful urbanism. It expanded on the past rather than destroying it [19]. It struck a balance between architectural drama and conceptual substance rather than overspending on spectacle. The end product is a cultural hub for Cape Town that appeals to both local and international audiences, serving as a potent example of how adaptive reuse can transform cities and structures.

5.4.1 Living with the Silo: Everyday Use and Public Reception

The Zeitz MOCAA has garnered a lot of attention since its September 2017 debut, both as an architectural landmark and as a cultural organization that aims to change the way that people across the world view African art. Regular use and public opinion have shown out to be key indicators of its legitimacy. The impact of the museum is gauged not only by visitor numbers and accolades but also by the experiences of local communities, artists, and critics navigating its place in Cape Town's waterfront environment.

In the museum's early years, hundreds of thousands of individuals are said to have visited in the first 12 months, which is an astonishing figure. A sizable section of this audience consisted of foreign visitors, who were already lured to the V&A Waterfront as Africa's most popular destination [1, 2]. Their presence contributed to Zeitz MOCAA's reputation as a must-see location, frequently likened to cultural flagships like the Guggenheim Bilbao or the Tate Modern in London [20]. However, the share of local visitors has been smaller, indicating the social and economic restrictions that determine museum access in South Africa, even while the influx of foreign visitors has increased revenue and global prominence. The cost of admission is still too high for a large number of Capetonians, particularly those from low-income neighbourhoods on the outskirts of the city.

Both adoration and ambivalence can be seen in the comments made by African artists and commentators. Many saw the establishment of a museum specializing in modern African art on African territory as a long-overdue remedy for centuries of exclusion. African artists have had access to unmatched visibility through Zeitz MOCAA exhibitions, which have exposed their work to a global audience and validated African innovation in international cultural circles [1]. The museum has held historic exhibitions that have won praise from critics, such as retrospectives of well-known African artists. However, some critics contend that rather than continuously promoting up-and-coming or marginalized voices, the institution runs the risk of perpetuating hierarchies by giving preference to artists who are already well-known in Western markets. The larger conflict between creating a top-tier organization and acting as a grassroots forum for cultural empowerment is reflected in these discussions.

The museum has received numerous awards. Its adaptive reuse of the silo building, which is frequently hailed as a triumph of architectural ingenuity, has earned it attention and accolades on a global scale. The structure itself has been hailed as one of the most remarkable cultural venues on the planet because of its sculptural galleries and vast atrium. The edifice serves as both a museum and a gigantic work of art in and of itself, and for many tourists, it is worth the admission fee alone.

However, criticisms of elitism and accessibility continue to coexist with the admiration. In a highly unequal nation where a large portion of the populace cannot readily afford cultural consumption at international tourist rates, admission fees continue to be a topic of contention. Some contend that the museum's placement within the commercialized, tourist-heavy V&A Waterfront further distances it from the local population. Despite its pan-African symbolism, some argue that the museum functions more as a global exhibition than as a deeply ingrained communal institution [21]. These critiques highlight how challenging it is to strike a balance between diversity and financial viability.

These dualities are reflected in the museum's daily operations. Zeitz MOCAA is crowded with tourists from around the world, students on guided tours, and art enthusiasts taking in temporary exhibits. On the other hand, it finds it difficult to completely democratize access for the groups that most require public space and cultural validation. The way it was received demonstrates the possibilities and constraints of cultural mega-projects in unequal society. In the end, Zeitz MOCAA has been successful in making a name for itself in the cultural world and solidifying African art's position there [1] [2]. However, its conflicting reactions serve as a reminder that cultural organizations must constantly walk a tightrope between local relevance and global ambition.

5.5 Beyond the Atrium: Impact, Contradictions, and Lessons of Zeitz MOCAA

A bold adaptive reuse of a former industrial silo, the first large-scale institution dedicated to modern African art on the continent, and a symbolic turning point in Cape Town's worldwide cultural standing, the Zeitz MOCAA is often referred to as a cultural game-changer [1]. However, its accomplishments must be evaluated in light of the inconsistencies and complications that come with any undertaking of this size. A broader picture that combines successes with warnings for adaptive reuse in the Global South is shown by looking at quantitative results, qualitative impacts, environmental aspects, externalities, and transferability.

Quantitatively speaking, the museum has done well as a destination for tourists. It received about half a million visitors in its first year, a number that was supported by Cape Town's status as a major tourist destination. In addition to hosting scores of exhibitions that unite established and up-and-coming artists from over the continent, it has generated jobs in the construction, operations, and cultural programming sectors. Retail businesses, events, and ticket sales have produced the revenue streams required to augment institutional and philanthropic assistance. These quantifiable benefits show how well the museum can integrate with the V&A Waterfront's cultural and commercial life [5].

Perhaps even more significant are the qualitative effects. After decades where recognition was mostly granted by European or American organizations, Zeitz MOCAA has provided African contemporary art with an institutional home on African territory, marking a symbolic change [1]. According to curators and artists, the museum has given African voices credibility and increased their visibility in international cultural circles. Additionally, it has changed Cape Town's reputation as a city with top-notch cultural infrastructure as well as a tourism destination with picturesque scenery. The idea that history can be used to serve new forms of public value is embodied by the silo's transformation from an industrial warehouse to a cultural landmark.

Environmental effects are important, although being less well-known. The project avoided the emissions linked to new construction by reusing the silo instead of demolishing it, saving resources and embodied carbon. Although critics point out that the museum should have done more to incorporate sustainability measures, incremental greening—natural light optimization, some landscaping—adds modest environmental benefits to the architecture. Nonetheless, the preservation of a significant industrial building is a precedent for adaptive reuse and has significance in the global discussion on sustainable design.

This image is complicated by the externalities of the endeavour. The addition of a well-known cultural organization has strengthened the already gentrified and commercialized nature of Cape Town's V&A Waterfront [22]. Inequalities between the coastline and outlying townships, where cultural investment is limited, could be made worse by rising property values and more tourist traffic. Accessibility is still a problem because of the high cost of tickets and the layout of the Waterfront, which keeps the communities that need cultural inclusion the most on the outside. The conflict between local relevance and global ambition is emphasized by accusations of elitism and an excessive dependence on foreign tourists.

Transferability and longevity complete the evaluation. Zeitz MOCAA has demonstrated resilience by withstanding changes in leadership, intense criticism, and the COVID-19 pandemic's interruptions. Other Global South towns thinking about adaptive reuse of industrial legacy can learn from its hybrid governance style and varied revenue streams. Transferability is conditional, though, as not many cities have access to international philanthropy or a vacant silo in a tourist-heavy area. Replicating form is not as important as embracing principle, which includes preserving embodied resources, converting industrial artifacts into civic assets, and creating institutions that put cultural sovereignty first.

In conclusion, Zeitz MOCAA is a provocation as well as a success. In addition to highlighting the enduring problems of equity, accessibility, and reliance on global finance, it shows how adaptive reuse may be used to redefine history and establish cultural validity. Its paradoxes are instructive: resourceful urbanism focuses on overcoming obstacles, resolving conflicts, and generating public value in unexpected locations rather than on flawless results. The historic silo's carved atrium gives us a sense of the endeavour's strength and vulnerability, serving as a symbol of how African towns may reimagine their histories to forge more inclusive cultural futures.

References

1. Nielsen V (2023) Ambivalent art at the tip of a continent: the Zeitz MOCAA and its quest for global recognition. In: Global art in local art worlds. Routledge, New York, pp 77–99
2. De Villiers RET (2016) The V&a Waterfront as workplace and leisure space for Capetonians. Doctoral dissertation, Stellenbosch: Stellenbosch University
3. Sulcas R (2023) Can she revive the largest museum on the African continent?. International New York Times, NA-NA
4. Ferreira S, Visser G (2007, September). Creating an African Riviera: revisiting the impact of the Victoria and Alfred waterfront development in Cape Town. In: Urban Forum (vol 18, No. 3). Springer Netherlands, Dordrecht, pp 227–246
5. Mokobane MM (2023) Understanding the relationships between tourism, and heritage conservation at the Victoria & Alfred Waterfront. Available from https://open.uct.ac.za/items/4993a57f-155b-4194-935c-da67f51f39d3

6. Kilian D, Belinda Dodson DR (1995) The capital see-saw: understanding the rationale for the Victoria and Alfred redevelopment. S Afr Geogr J 77(1):12–20
7. Morezzi E (2014) Memory and transformation: the case of Victoria & Alfred Waterfront, Cape Town. PORTUS PLUS 4:1–10
8. De Tolly P (2009) Cape Town's central waterfront: past, present, future: challenge and change. Hist Environ 22(2):22–28
9. Bickford-Smith V (2009) Creating a city of the tourist imagination: The case of Cape Town, 'The fairest Cape of them all'. Urban Stud 46(9):1763–1785
10. Rowlands M (2023) Recognition. In: Global art in local art worlds. Routledge, New York and London, pp 100–105
11. Heatherwick T (2015) Thomas heatherwick: making. The Monacelli Press, LLC
12. Supple B (2024) The Zeitz MoCAA museum. Available from https://open.bu.edu/items/091f2c5a-4e74-405f-b9ad-224047551cab4
13. Cheshmehzangi A (2020) Reflecting on urban identities. In: Identity of cities and City of identities. Springer Singapore, Singapore, pp 133–178
14. Roos A (2017) Swiss sensibility: the culture of architecture in Switzerland. Birkhäuser
15. Cheshmehzangi A, You N, Siri J, Birch E (2024) Western cape industrial symbiosis programme (WISP) delivered by GreenCape in the City of Cape Town, South Africa. In: Harnessing urban innovation to unlock the sustainable development goals. Springer Nature Singapore, Singapore, pp 133–138
16. Zorloni A, Resch M (2016). Turning collections into philanthropic opportunities. In: Art wealth management: managing private art collections. Springer International Publishing, Cham, pp 1–18
17. Pollio A (2018) Social entrepreneurship, and the technopolitics of millennial development in Cape Town. Available from https://researchers.westernsydney.edu.au/files/94852977/uws_71067.pdf
18. Swart L, Swilling M, Gcanga A (2024) Exploring a water–energy–food (WEF) nexus approach to governance: a case study of the V&a waterfront in Cape Town, South Africa. Energies 17(16):4005
19. Saldarriaga Restrepo C (2023). Nearby materials: wood and earth as nearby materials for the redevelopment of Zeitz. Master's thesis
20. Evans G (2003) Hard-branding the cultural city–from Prado to Prada. Int J Urban Reg Res 27(2):417–440
21. Inggs A (2014) African city-Cape Town in pieces/aesthetics, theories, narratives, fragments. Available from https://open.uct.ac.za/handle/11427/6860
22. Ryce-Paul R (2004) Waterfront revitalisation: profitability vs social equity strategies shaping the urban waterfront. Master Thesis at Columbia University, May 2024

Part III
Circular Resource Management

Chapter 6
Loops of Value: The Kalundborg Industrial Symbiosis and the Architecture of Circular Exchange in Denmark

6.1 From Linear Industry to Circular Synergy: Context and Political Economy of Kalundborg

The tale of Kalundborg's industrial symbiosis is based on the complex interactions of location, industrial necessity, and political economy rather than on abrupt technological advancements [1]. It is crucial to place the origins of this small Danish town in the context of Denmark's industrialization trajectory in the middle of the twentieth century, the local resource conditions, and the structural constraints of traditional production models in order to comprehend why it became the most cited example of circular industrial collaboration in the world.

Although not as rapidly as many of its European neighbours, Denmark was rapidly industrializing and urbanizing in the post-war era. Through manufacturing, medicines, energy, and marine trade, the nation's economy was shifting from an agrarian foundation to a more integrated one with international markets. However, there was a relative lack of resources when this industrialization took place [2]. Due to its limited coal reserves, Denmark was mostly dependent on imported fuels, especially oil, to power its expanding industrial base. Because of this reliance, energy prices rose, and resource efficiency became a national priority. Although these vulnerabilities would later be exacerbated by the oil crises of the 1970s, Danish companies and politicians were well aware of the importance of prudent resource management as early as the 1950s and 1960s.

Situated on Zealand's west coast, Kalundborg itself presented both opportunities and limitations that influenced its industrial growth. It was an appealing location for energy-intensive and export-focused enterprises because it was a coastal town with access to deepwater ports. In addition to Novo Nordisk's pharmaceutical and enzyme plants, as well as other chemical and manufacturing facilities, the Statoil (now Equinor) refinery and the Asnaes Power Station were positioned advantageously there [3]. Kalundborg was a logical industrial cluster because of its

A. Cheshmehzangi, *Resourceful Urbanism*, Urban Sustainability,
https://doi.org/10.1007/978-981-95-4138-6_6

closeness to Copenhagen and other markets, somewhat cheap land costs, and salt-water availability for cooling.

Yet, there were dangers associated with this grouping. At first, the industries functioned separately and in silos with minimal collaboration. Each produced waste streams that were either released into the environment or required expensive treatment due to their high energy and water consumption [4]. For example, the Asnaes Power Station released waste steam and excess heat from burning coal to produce heat and electricity. Wastewater, sulfur, and other byproducts from the refinery were produced, and their disposal was expensive. Biomass sludge from Novo Nordisk's operations needs to be removed. Businesses and the town were burdened financially and environmentally by these outputs, which were viewed as liabilities.

Kalundborg was a prime example of the linear industrial model during this early stage: resources were taken, turned into goods, and waste was thrown away. The prevailing worldwide paradigm of the middle of the twentieth century, which placed a higher priority on economic expansion than environmental conservation, was reflected in this model. However, the effects were most noticeable in places like Kalundborg [5]. Carbon-intensive manufacturing systems were produced by a heavy reliance on fossil fuels. Waste contributes to environmental deterioration and health issues by releasing contaminated air and water. While the municipality grappled with infrastructure issues including district heating, wastewater treatment, and the escalating costs of resource delivery, the town's inhabitants were exposed to the byproducts of industrial growth.

These circumstances paved the way for a slow but significant rethinking of industrial organization. Companies and municipal governments started to consider whether by-products may be recycled as inputs elsewhere instead of considering garbage as an unavoidable end point. This change was initially articulated in terms of practical cost and necessity calculations rather than the buzzwords of "circular economy" or "industrial ecology," which would only become popular decades later. Sludge, gypsum, steam, and waste heat were all acknowledged as possible resources rather than worthless waste [6]. This logic of reliance and scarcity developed Kalundborg's political economy as businesses realized that working together could lower operating costs, satisfy legal obligations, and lessen the municipality's infrastructure providing load.

Therefore, the background of Kalundborg's industrial symbiosis demonstrates how several forces came together: Denmark's reliance on imported energy, the spatial concentration of various industries around a coastal town, the financial and environmental difficulties of linear industrial systems, and the realization that isolated operations were unsustainable [7]. These limitations gave rise to the seeds of a new paradigm, one in which businesses may prosper by connecting with one another in networks of exchange rather than by isolating themselves.

The history of Kalundborg serves as a reminder that innovative urbanism at the industrial level involves redefining issues rather than focusing on big plans [8, 9]. The traditional linear production model—extract, use, discard—turned out to be extremely expensive and harmful. On the other hand, Kalundborg's initial attempts at circular synergy demonstrated that collaboration, not just competition, might create new kinds of public and private value and unleash efficiencies. This historical, political, and economic context clarifies why Kalundborg became the quintessential example of industrial symbiosis—not because it aimed to build a model city, but rather because geography and necessity made creative cooperation the most sensible course of action.

6.2 From Deals to Networks: The Emergence of Symbiosis in Kalundborg

It took time for Kalundborg's industrial symbiosis to emerge as a carefully thought-out invention. Instead, it developed gradually as a result of practical agreements between nearby industries that gradually came together to form a well-coordinated network of transactions. Over the course of three decades, what started out as bilateral agreements to reduce waste and save money developed into an internationally recognized system of industrial partnership [10]. Thus, the development of symbiosis in Kalundborg is less a story of innovative design and more an investigation of how systemic change may be produced via trust, necessity, and small-scale collaborations.

When the Asnaes Power Station and the Statoil (later Equinor) refinery looked for ways to save operating costs in the 1960s, the first signs of cooperation appeared [1]. Both facilities used a lot of resources, generating waste heat and steam and using a lot of water for cooling. The refinery started buying extra steam from the power plant rather than using separate, ineffective systems. This agreement gave the power plant a steady source of income for what had previously been a lost by-product while lowering expenses and emissions by reducing the refinery's requirement to produce its own steam. This first deal's success set a precedent: if the proper partner was prepared to use it, what had been viewed as waste might turn into a resource.

Exchanges grew from this humble beginning. The excess heat from the power plant was diverted into a district heating system that serves the town of Kalundborg after being released into the air or water [11, 12]. Households benefited from reduced heating expenses and increased energy security, and the power plant benefited from favourable public perception and favourable regulations [13, 14]. Later, gypsum was created as a by-product of the power plant's desulfurization operations and given to a cement producer. Similar to this, Novo Nordisk produced sludge as a result of its operations at its sizable pharmaceutical and enzyme manufacturing facility. This sludge was transferred to nearby farmers, who used it as fertilizer, establishing an agricultural link in the symbiotic chain rather than treating it as garbage.

This intricate network of resource movements had come to define Kalundborg's industrial landscape by the 1980s and 1990s. Before being treated and re-distributed for additional industrial or municipal uses, the water was pumped through pipelines several times for cooling purposes [5]. Chemical makers received sulphur that was removed from refinery processes. Waste streams were continuously reframed as inputs, resulting in a cascade of utility that extended across the municipal, industrial, and agricultural spheres.

Both technical compatibility and institutional innovation contributed to the longevity of these collaborations. Formal contracts made it clear who was responsible for infrastructure investments like pipelines and treatment plants and guaranteed supply reliability. The unofficial culture of mutual respect and trust that emerged between businesses, industries, and the municipality, however, was just as significant [1]. Given Kalundborg's small size, the close-knit bonds between managers, engineers, and public servants encouraged a sense of collective accountability for the welfare of the city [15]. This embeddedness meant that disputes could be resolved more easily, and the risks of opportunism were reduced.

A wide range of interests are represented by the main players in this process. The massive pharmaceutical company Novo Nordisk was motivated to improve its sustainability profile and handle its byproducts properly. As a significant energy producer, Statoil/Equinor wanted to increase efficiency and adhere to stricter environmental standards. As the biggest energy facility in the area, the Asnaes Power Station served as the hub, supplying several partners with heat, steam, and byproducts [16, 17]. The municipal utility made sure that exchanges matched public demands, such district heating and water supply, by providing both regulatory control and infrastructure assistance [18]. Despite being lesser participants, farmers played a crucial role in connecting the local agricultural economy to the industrial environment by receiving biomass by-products.

The transition from opportunistic transactions to an organized symbiotic system was signalled by the institutionalization of these exchanges [4]. While collaborative investments in distribution systems, treatment facilities, and pipelines physically incorporated the network into the landscape, contracts supported the exchanges' financial stability. This architecture gradually locked in cooperation, strengthening the network's resilience and reducing its reliance on personal connections. By the 1990s, Kalundborg had evolved from a series of bilateral agreements [4, 8] to a fully functional system of industrial symbiosis that caught the interest of academics, decision-makers, and environmentalists worldwide.

The development of the Kalundborg model serves as an example of how resourcefulness and incrementalism can lead to systemic transformation. The network was not entirely planned by a single actor. Rather, each transaction addressed pressing issues, such as cutting expenses, adhering to rules, or maximizing the value of waste. These practical solutions eventually came together to form a cohesive circularity system, proving that inventive urbanism at the industrial scale can be constructed from the ground up.

The early history of Kalundborg teaches us the value of taking baby steps. The municipality established a resilient circularity model that continues to influence discussions about sustainable industry globally by redefining trash as opportunity, fostering trust across institutional borders, and layering exchanges over decades.

6.2.1 From Ad-Hoc Deals to Structured Network

The best way to interpret the history of Kalundborg's industrial symbiosis is as a progression: from practical bilateral agreements in the 1960s to a systematic, globally accepted system of circular exchange by the 1990s. Technical innovation, infrastructure spending, and the expanding role of local government were the main forces behind this shift from improvisation to institutionalization [4]. By the time international observers started looking into Kalundborg, the town had evolved from a collection of companies to a model for how resource loops may be set up across scales and sectors.

The initial steps were small. In order to prevent unnecessary venting and lower the refinery's operating expenses, the Asnaes Power Station consented to send excess steam to the Statoil refinery in the 1960s [16]. Soon after, the power plant's waste heat was converted into a district heating system, giving Kalundborg homes reasonably priced warmth. Ad hoc in nature, these early agreements were driven more by technical viability and cost reduction than by a comprehensive sustainability concept. However, they did pioneer the idea that shared value could be created from byproducts.

New exchanges were added to the original partnerships over the 1970s and 1980s. Cement producers received the gypsum generated by the power plant's flue-gas desulfurization process. The refinery's extracted sulphur was diverted to the chemical industry. Novo Nordisk started providing local farmers with its nutrient-rich sewage, which they utilized as fertilizer. Pipeline-circulated water was cleansed, reused, and redistributed several times, increasing the usefulness of a limited resource. Along with new agreements about cost-sharing and liability, each addition necessitated new technical and infrastructure solutions, including as pipelines, pumping stations, and treatment facilities. By the late 1980s, Kalundborg's network had evolved into a self-reinforcing system of resource flows [4, 17] rather than a patchwork of transactions.

The network's formalization was supported by investments in infrastructure that physically secured transactions. Once water or steam pipelines were established, they established strong connections that promoted sustained cooperation. Innovation in technology also contributed. For instance, desulfurization technology produced useful byproducts in addition to lowering emissions, while improvements in wastewater treatment made multi-cycle water use feasible. These technological changes expanded the network by generating additional "exchangeables" that businesses could distribute.

In this growth, municipal governance played a crucial role. The municipality of Kalundborg made it easier for businesses to coordinate, especially in areas where industrial flows and public infrastructure, like district heating, met. By providing regulatory monitoring, it made sure that exchanges complied with municipal requirements and environmental norms. The municipality served as a trust broker, which was equally significant. Personal ties between engineers, managers, and public authorities in a small town lowered opportunistic risks and prompted businesses to collaborate on investments. The municipality's participation validated the system as benefiting the community at large as well as business interests.

Kalundborg had gained international recognition by the 1990s. Industrial ecology experts hailed it as the first operational instance of industrial symbiosis in history, a living test bed for circular economy and resource efficiency ideas. The town's concept was studied by policy delegations and industry executives, and it was frequently featured in sustainability reports and case studies. What had started out as a collection of economical improvisations was now hailed as a well-organized system of international importance.

The progression from steam pipes to systemic networks serves as an example of how little, localized interactions can build up to create revolutionary models. Although Kalundborg did not initially plan its symbiosis, it developed a structured network that still shapes the global discourse on industrial circularity via decades of practical problem-solving, infrastructure development, and municipal assistance.

6.3 The Invisible Handshake: Governance and Finance in Kalundborg's Symbiosis

In contrast to many well-known sustainability initiatives that depend on top-down planning, Kalundborg's industrial symbiosis has persisted for more than 50 years without having its conditions dictated by a central authority. Its unique governance and financial architecture—decentralized, based on reciprocal benefits, supported by infrastructure, and bolstered by trust—is what gives it its longevity [19]. This hybrid paradigm, which falls in between unofficial handshakes and formal contracts, provides insightful guidance on how to institutionalize collaboration without sacrificing adaptability.

Instead of being a hierarchical organization, Kalundborg's governance is best defined as a decentralized network [20, 21]. The symbiosis is not owned or controlled by one body. Rather, it is made up of bilateral and multilateral agreements between businesses that are mediated by contractual duties and shared infrastructure. While each company continues to act independently and pursue its own business goals, they all take part in transactions that save costs and improve sustainability. Resilience is made possible by this decentralization; even if one exchange stops, others can go on without the system as a whole failing.

On the other hand, the network is stable because of contracts and infrastructure [21, 22]. Agreements outline the amounts, costs, and quality requirements for flows like sludge, water, gypsum, and steam. These agreements, which frequently last for decades, offer long-term stability and support the capital expenditures necessary to construct pipelines and treatment facilities. Because steam pipelines and water recycling systems are too expensive to quickly abandon and rely on continuous cooperation to be valuable, the infrastructure itself locks in cooperation once it is constructed. Physical infrastructure thus turns into a tool for governance, integrating collaboration into the environment.

The financing methodology has been practical and cost-sharing. Every time a new exchange is suggested, the companies concerned bargain about how to divide the infrastructure investment. For instance, both parties split expenses based on anticipated benefits if excess steam from the Asnaes Power Station is fed to Novo Nordisk. Instead than relying on nebulous concepts of corporate responsibility, these estimates are based on tangible benefits, such as decreased trash fees and energy expenditures [23]. The system's longevity has been largely attributed to this financial pragmatism. Every agreement had to be profitable for all parties, guaranteeing that financial incentives and environmental advantages matched.

Thus, the financial foundation of Kalundborg's symbiosis is mutual gain. By lowering operating expenses and producing new sources of income, exchanges established a win-win scenario that promoted involvement. While nearby residents enjoyed reasonably priced district heating, the Asnaes Power Station made money by selling excess heat [16]. Farmers were able to obtain fertilizer, while Novo Nordisk was able to reduce the expenses associated with disposing of garbage. These mutually beneficial effects created local legitimacy and strengthened the social compact of the system, benefitting not only the businesses but also the larger community.

Trust and transparency processes are the softer aspects of governance. In the comparatively small town of Kalundborg, managers, engineers, and local authorities are acquainted with one another [23]. This proximity fosters a culture of reciprocity and informal problem-solving. Businesses are more inclined to work together to overcome technological problems in a pipeline or exchange than to resort to legal action. It has also been crucial to provide information regarding resource flows and operational requirements in a transparent manner. Although formal safeguards are provided by contracts, cooperation is sustained over decades by enduring connections and local community integration.

In this governance structure, the municipality has a minor but important function. It facilitates, regulates, and benefits the symbiosis without having any control over it. For example, the district heating system connects industrial byproducts to public welfare by directly serving nearby houses [4]. Exchanges comply with regulations thanks to municipal supervision of water use and environmental norms. The municipality strengthens the legitimacy and robustness of the symbiosis by tying industry flows to community needs and establishing it within a civic framework.

The governance and financial model of Kalundborg exemplifies a type of what could be referred to as the "invisible handshake": collaboration maintained by a combination of agreements, infrastructure, trust, and reciprocal benefit rather than centralized authority or altruistic goal [4]. It questions the notion that central planning or large-scale designs are necessary for sustainability. Rather, it demonstrates how embedded practices and practical incentives can help decentralized networks institutionalize collaboration.

The lesson for ingenious urbanization is obvious. Formal institutions are not always necessary for governance, and substantial subsidies are not always necessary for finance. The most resilient systems can occasionally be formed through small agreements based on shared interests that are progressively formalized through relationships and infrastructure. Kalundborg shows that cooperation can be successful and long-lasting even in the cutthroat world of business, as long as the framework of cost-sharing and trust is properly developed.

6.4 Pipes and Partnerships: Implementation, Flows, and Outcomes in Kalundborg

The Kalundborg Industrial Symbiosis, a network of exchanges where trash from one company becomes resource for another, is frequently praised for its elegance. However, the system's physical implementation—the pipelines, pumps, treatment facilities, and contractual agreements that made the concept a reality—is what really sets it apart [21]. Kalundborg has developed a highly integrated system of resource loops over many years that yield quantifiable social and environmental advantages. Its narrative focuses more on the tangible reality of technological flows than it does on abstract models.

Heat and steam exchanges are at the center of the system. Once a traditional coal-fired plant, the Asnaes Power Station became the focal point of the network. It directed these flows to industrial neighbours rather than releasing waste heat or surplus steam. Early recipients were Novo Nordisk and the Statoil/Equinor refinery, which used steam in their manufacturing operations [24]. Excess heat from industries was redirected into a district heating system that provided heat to thousands of nearby homes. This one transaction significantly decreased the community's need on individual oil-fired boilers, lowering pollution and supplying reasonably priced heating (See Table 6.1).

Another important pillar was water management. Kalundborg's industrial production put a strain on the region's freshwater supplies by using large amounts of cooling and process water. In order to move water between businesses and reuse it several times before undergoing final treatment, a system of pipelines was built. In order to protect aquifers and guarantee long-term water security, surface water and treated wastewater took the role of groundwater exploitation. This multi-cycle water use is a prime example of the cascading utility principle, which states that value should be maximized before disposal.

The network was further diversified by by-product exchanges. The power station's desulfurization process yielded gypsum, which was then used as a raw ingredient to a nearby cement factory. Refinery-recovered sulphur was used into chemical production streams, generating extra income while minimizing disposal expenses [4, 5]. Novo Nordisk produced nutrient-rich sludge from its pharmaceutical and enzyme operations. The connection between industry and agriculture was closed when the sludge was sent to nearby farmers as fertilizer rather than being treated as garbage. For any of these flows to be used across industries, investments in storage, transportation, and quality control were necessary in addition to technical viability.

Table 6.1 Summary of principal resource exchanges, the pipes/plants that make them work, firm-level and public value captured, and the governance/finance mechanisms that stabilized each loop

Exchange/ Flow	Source Facility	Recipient/End Use	Enabling Infrastructure/ Technology	Firm Value Captured	Public Value Captured	Governance/Finance Mechanism
Process steam	Asnæs Power Station	Statoil/Equinor refinery; later other industrial users	Steam pipelines; metering and pressure control	Monetises a by-product; lowers venting; stable off-take revenue	Reduced local emissions from duplicate boilers	Long-term bilateral contract; capex cost-sharing proportional to benefit
Waste heat → district heating	Asnæs and other plants	Households and municipal buildings (district heating grid)	Heat exchangers; insulated transmission mains; load balancing	Reputation gains; predictable heat sales	Lower bills; cleaner air; winter energy security	Concession with municipal utility; tariff regulation; guaranteed minimum off-take
Cascaded water re-use	Multiple plants (cooling/process)	Reused across plants → treatment → municipal/industrial use	Raw/treated water mains; on-site polishing; reuse standards	Cuts intake and discharge fees; resilience in dry periods	Aquifer protection; reduced treatment burden	Multi-party service agreements; quality SLAs; shared O&M
FGD gypsum	Asnæs flue-gas desulfurization	Cement/board manufacturer	Enclosed conveyors; covered storage; moisture control	Avoided disposal; by-product revenue stream	Avoided quarrying; landfill diversion	Supply contract indexed to quality/moisture; take-or-pay volumes
Recovered sulfur	Refinery desulfurization	Chemical manufacturers	Sulfur recovery, filtration, bulk handling	Waste-to-product margin; compliance benefit	Fewer sulfur emissions; circular input for industry	Product-grade spec contract; logistics cost-sharing
Biomass sludge (biosolids)	Novo Nordisk (pharma/ enzymes)	Local farms (soil amendment)	Dewatering; hygienisation; agricultural spreading	Avoided disposal costs; social licence	Lower input costs for farmers; soil carbon co-benefits	Tripartite MoU (firm–farmers–municipality); liability and monitoring clauses

Notes: This table synthesizes material throughout 6.1–6.4, especially early steam/heat deals, water cascades, and by-product valorisation based on multiple references for the case study [1–6, 9, 11, 16–18, 20–23, 24]

These exchanges are supported by a vast infrastructure. Kalundborg is criss-crossed with kilometers of pipelines that connect facilities to municipal systems and to each other [16]. In order to achieve quality standards, treatment facilities process water and its byproducts. Industrial flows are integrated into public services like heating through utility connections. This infrastructure is the physical framework of the symbiosis, connecting businesses and integrating cooperation into the town's design. It is more than just plumbing.

The quantifiable results are significant. According to studies, the symbiosis has lowered CO_2 emissions by hundreds of thousands of tons annually, mostly as a result of a decrease in the need for industrial energy and a dependence on fossil fuels for heating [3]. With multi-cycle use lowering overall demand by millions of cubic meters per year, water savings have been as noteworthy. By-product swaps that divert solid waste from landfills result in additional emissions that are avoided. These economic advantages are equal to the environmental ones: homes gain from cheaper, cleaner heating, while businesses save millions of euros annually in avoided expenses.

Programmatic results are not limited to the industry. Residents' energy equity has increased thanks to the district heating system, and the town-country relationship has been reinforced by the use of industrial sludge in agriculture. As a result, the symbiosis extends into everyday life and strengthens the network's legitimacy by providing observable advantages to the public [3]. In contrast to many corporate sustainability initiatives, Kalundborg's results—from warm houses to fertilized fields—are apparent to its residents.

Although incremental, integration into larger planning frameworks has been important. The symbiosis is currently seen as a strategic asset by local and regional authorities, who align it with Denmark's national energy and climate goals. As an example of how industrial ecology and public policy may coexist [1], what started out as local improvisation has evolved into a recognized element of city-regional development.

The execution of Kalundborg emphasizes that in addition to goodwill, resourceful urbanization needs hardware. In the absence of the expenditure on pumps, plants, and pipes, symbiotic concepts would have remained theoretical [1]. The town has demonstrated how small technological exchanges can develop into a strong system with regional and global ramifications by establishing cooperation in physical flows.

6.4.1 Symbiosis in Daily Life: Local Benefits and Global Perceptions

The most revealing effects of the Kalundborg symbiosis may be observed in daily life, even though it is frequently praised in scholarly articles and sustainability reports. The effectiveness of industrial exchanges as well as the ways in which these flows interact with workers, households, and the town's larger character are what set this scenario apart [16]. The reputation of the approach has been firmly established by public opinion, both domestically and internationally.

District heating is the most obvious advantage for local residents. In order to lessen reliance on expensive oil-fired boilers, excess heat from the Asnaes Power Station and other facilities is pumped straight into residences. Residents benefit from increased security and less energy costs, especially during the severe winter months. Compared to separate heating units, the system offers additional environmental benefits, such as improved air quality due to decreased emissions. An uncommon and potent example of industrial ecology working for the public good is the direct connection between home comfort and industrial byproducts.

Residents and employees alike see changes in the quality of the environment. Pollution burdens have decreased as a result of less waste releases into nearby air and waters. Although these exchanges are still strictly regulated to prevent health hazards, farmers that use fertilizer made from Novo Nordisk's by-products report cost savings and increased yields. Employees in participating industries take delight in helping to create a system that is admired around the world, which strengthens their feeling of place. Kalundborg is now seen as a leader in sustainability rather than just an industrial town.

This reputation has been enhanced by international recognition. Kalundborg is frequently highlighted in sustainability conferences, journals, and exhibitions, and it has won honors. On study tours, groups of academics, legislators, and business executives come to the town to observe the system in action. Kalundborg is now seen as a model for circular economy activities as a result of the pressure and prestige this focus has brought about.

Yet critiques persist. Due to limited public access to data on the specific environmental implications of exchanges, transparency has occasionally been questioned. Many symbiotic flows still come from fossil fuel-based businesses, according to critics, which raises concerns about their long-term compatibility with Denmark's renewable energy targets. Vulnerability arises from reliance on heavy industry; the network may lose important nodes if core companies, like the refinery, downscale or relocate. By reminding spectators that symbiosis must always adjust to new energy and economic conditions, these worries moderate the story of success.

All things considered, daily operations and public opinion portray Kalundborg as a lived reality and a universal emblem. In addition to providing real benefits to workers and households, it also raises important questions regarding its reliance on legacy industries. Making industrial cooperation visible and valuable at all scales—from the comfort of a local home to the prestige of international recognition—is what makes it successful, not perfection.

6.5 Loops and Limits: Evaluating the Impacts of Kalundborg's Symbiosis

Kalundborg's industrial symbiosis has long been referenced in sustainability manifestos, policy briefs, and textbooks as the model of resource efficiency. However, one must balance its quantifiable effects against its transferability, externalities, and vulnerabilities in order to completely understand its relevance. Unquestionably amazing, the model's lessons are best understood when viewed with both caution and admiration.

Kalundborg's accomplishments are noteworthy from a quantitative standpoint. According to studies, the network prevents hundreds of thousands of tons of CO_2 emissions every year, mostly by replacing fossil fuels with heat and steam exchanges. By conserving millions of cubic meters of freshwater annually, multi-cycle water use helps to safeguard nearby aquifers. Large amounts of materials, such as gypsum, sulphur, and biomass sludge, are diverted from waste streams into useful applications through by-product exchanges. According to economic assessments, by avoiding disposal costs, lowering energy bills, and creating new revenue streams, participants together save tens of millions of euros every year. Additionally, thousands of households benefit from district heating, which is less expensive than traditional systems could provide. These measurements provide concrete proof that symbiosis is materially transforming rather than just metaphorical.

Qualitative effects provide more context. As a leader in sustainability, Kalundborg has assimilated into Denmark's national identity, enhancing its standing as a green economy driven by innovation. The network has given local businesses legitimacy and confidence, proving that environmental cooperation and competitiveness can coexist. In addition to cheaper heating, residents take satisfaction in living in a neighbourhood that is well-known throughout the world. The town's name has come to represent "industrial symbiosis," solidifying its position in the conversation about sustainability on a worldwide scale.

However, it is impossible to ignore externalities and vulnerabilities. A large portion of the network is still connected to fossil fuel-based businesses, especially the refinery and the coal-fired power plant that served as key hubs for a long time. As Denmark moves forward with its green transition, a system intended to reuse fossil byproducts may struggle due to this dependence, which increases the risk of "lock-in." Renewable energy sources may lower total emissions, but they also cut off the supply of steam, heat, and byproducts that support the network. Changes in the market, including the closure of important facilities or downsizing, could cause the system to become unstable. Despite its efficiency, critics contend that Kalundborg's model nevertheless represents a compromise by prolonging the existence of polluting enterprises through the valuation of their waste.

However, one of Kalundborg's biggest advantages has been its longevity. Over the course of more than 50 years, the system has changed to accommodate new laws, advancements in technology, and demands from the market. Its decentralized governance, contractual stability, and ingrained trust among local actors are what give it resilience. Kalundborg has survived because it combines financial incentives with environmental objectives, in contrast to many pilot programs that fail when funding runs out. Although its history indicates that it can be reinvented, its ability to adapt to the green transition is still up for debate.

Regarding transferability, Kalundborg provides guidance as well as inspiration. It has served as a model for circular industrial development and has influenced eco-industrial parks all across the world, from North America to China. However, because Kalundborg's foundational elements—the close proximity of many businesses, a small-town culture of trust, and helpful municipal governance—are hard to duplicate elsewhere, replication has frequently failed. Despite being intangible, trust and embeddedness are just as important as pipes and contracts.

In conclusion, Kalundborg's symbiosis is a paradox as well as a standard. It demonstrates how industry collaboration may result in substantial economic and environmental benefits by converting trash into resources on a large scale. However, it also serves as a reminder that circular systems are context-specific, influenced by historical trends, institutions, and geographic location. The lesson is to apply Kalundborg's tenets—pragmatism, reciprocity, and incrementalism—wherever chances present themselves, rather than to completely replicate it. Its loops have changed a community, motivated people worldwide, and shown that sustainability can come from necessity and compromise rather than just ambition. On the other hand, as industrial landscapes change toward post-fossil futures, its limitations warn us to continue being flexible.

References

1. Ehrenfeld J, Gertler N (1997) Industrial ecology in practice: the evolution of interdependence at Kalundborg. J Ind Ecol 1(1):67–79
2. Chertow MR (2000) Industrial symbiosis: literature and taxonomy. Annu Rev Energy Environ 25:313–337
3. Jacobsen NB (2006) Industrial symbiosis in Kalundborg, Denmark: a quantitative assessment of economic and environmental aspects. J Ind Ecol 10(1–2):239–255
4. Chertow MR, Ehrenfeld J (2012) Organizing self-organizing systems: toward a theory of industrial symbiosis. J Ind Ecol 16(1):13–27
5. Boons F, Chertow M, Park J, Spekkink W, Shi H (2017) Industrial symbiosis dynamics and the problem of equivalence: proposal for a comparative framework. J Ind Ecol 21(4):938–952
6. Lombardi DR, Laybourn P (2012) Redefining industrial symbiosis: crossing academic–practitioner boundaries. J Ind Ecol 16(1):28–37
7. Paquin RL, Howard-Grenville J (2013) Blind dates and arranged marriages: longitudinal processes of network orchestration in industrial symbiosis. Org Stud 34(11):1621–1653
8. Doménech T, Davies M (2011) The role of embeddedness in industrial symbiosis networks: Phases in the evolution of industrial symbiosis. Bus Strat Environ 20(5):281–296
9. Jensen PD, Basson L, Hellawell EE, Bailey MR, Leach M (2011) Quantifying 'geographic proximity': experiences from the United Kingdom's national industrial symbiosis programme. Resources Conserv Recycling 55(7):703–712
10. Lund H (2007) Renewable energy strategies for sustainable development. Energy 32(6):912–919
11. Lind J, Rundgren E (2017) Industrial symbiosis in heat recovery collaborations between data centers and district heating and cooling companies
12. Butters C, Cheshmehzangi A (2017) Cities, climate and cooling. In: Designing cooler cities: Energy, cooling and urban form: The Asian perspective. Springer Singapore, Singapore, pp 5–19
13. Cheshmehzangi A, Butters C (2015, June) Refining the complex urban: the study of urban residential typologies for reduced future energy and climate impacts. In: Proceedings of 8th conference of the international forum on urbanism. Incheon, South Korea

14. Butters C, Cheshmehzangi A (2017) Integrated urban design: Study for a housing district in Ningbo, China. In Designing cooler cities: Energy, cooling and urban form: The Asian perspective. Springer Singapore, Singapore, pp 105–116
15. Keränen ML (2024) Qualitative evaluation of industrial symbiosis in a circular economy context: case study: the Kymenlaakso industrial cluster in Finland. Available from https://lutpub.lut.fi/handle/10024/167086
16. Liss L (2017) Industrial symbiosis: Kalundborg, Denmark. In: Global case studies: lessons for Ashdod from around the World, pp 28–33. Cambridge. Available from https://www.industrial-urbanism.com/_files/ugd/b1a909_d1d7c09833b54c59b6fef46d64e3d42d.pdf#page=28
17. Valero A, Usón S, Costa J (2012) Exergy analysis of the industrial symbiosis model in Kalundborg. Proc. ECOS, 25(218):1–218
18. Larsen M, Bech N, Bidstrup T, Christensen NP, Vangkilde-Pedersen T, Biede O (2007) Kalundborg case study, a feasibility study of CO2 storage in onshore saline aquifers. Danmarks og Gronlands Geologiske Undersogelse Rapport 2:79
19. Valentine SV (2016) Kalundborg Symbiosis: fostering progressive innovation in environmental networks. J Clean Prod 118:65–77
20. Domenech, Davies M (2011) Structure and morphology of industrial symbiosis networks: the case of Kalundborg. Procedia-Soc Behav Sci 10:79–89
21. Grann H (1997) The industrial symbiosis at Kalundborg, Denmark. The industrial green game. Implications for environmental design and management, 117–123
22. Branson R (2016) Re-constructing Kalundborg: the reality of bilateral symbiosis and other insights. J Cleaner Prod 112:4344–4352
23. Paché G (2024) Kalundborg industrial symbiosis: circular strategy in the light of mutualism. J Strat Innov Sustain 19(3):89–97
24. Pereira GS (2020) Identificação de práticas relevantes para o desenvolvimento de ecoparques industriais. Available from https://repositorio.unesp.br/server/api/core/bitstreams/77d1d97a-3769-404e-aaf9-ce92d632411f/content

Chapter 7
Everyday Circularity: Informal Labour and the SWaCH Model in Pune, India

7.1 Building Circularity from the Margins: Context and Political Economy of Pune's Waste System

The historical, demographic, and political-economic circumstances of Pune must be taken into consideration in order to comprehend the formation of the SWaCH cooperative and the larger Zero-Garbage initiatives in the city [1]. The trash management situation in Pune is representative of larger trends that are typical of fast urbanizing cities in the Global South: rapid population growth, poor municipal infrastructure, reliance on unorganized labour, and the ongoing failure of capital-intensive, centralized waste management plans. These forces gave rise to a unique kind of resourceful urbanism where local communities and marginalized labour reshaped the politics of garbage.

By the late twentieth century, Pune was one of India's fastest-growing metropolitan centres. By the 2000s, the city's population had grown from about 1.5 million in the 1980s to over 3 million, and forecasts indicate that this growth will continue. The development of significant IT parks, industrial growth, and the city's status as a center for education all contributed to this boom. The need for housing, transportation, water, and sanitation all increased as a result of the city's expansion. Specifically, solid waste management became a hot spot. Pune was estimated to produce over 1000 tonnes of municipal solid trash per day in the early 2000s, and this amount has only gone up since then. Like many other metropolitan administrations, the Pune Municipal Corporation (PMC) found it difficult to expand its systems in order to accommodate this demand [1–3].

Long before their contribution was acknowledged by any official institution, informal rubbish pickers played a crucial role in this vacuum. These labourers, who were mostly women from Dalit and other marginalized communities, searched landfills, trash cans, and streets for recyclables including metal, plastic, and paper. By diverting a sizable portion of rubbish from disposal sites, they supported their

A. Cheshmehzangi, *Resourceful Urbanism*, Urban Sustainability,
https://doi.org/10.1007/978-981-95-4138-6_7

own precarious incomes and supplied local recycling enterprises. However, the state did not see their labour, and society stigmatized them. Waste pickers endured intimidation, worked without safety gear, and received inconsistent pay based on changes in scrap values [4]. Nevertheless, as landfill crises grew, it became impossible to overlook the fact that they were the foundation of Pune's recycling sector.

Large amounts were hauled to locations like Uruli Devachi on the outside of the city, as the municipal system was mostly dependent on landfill disposal. These locations come to represent the shortcomings of traditional trash handling. They were overworked and badly designed, producing leachate that tainted groundwater, released methane, and caused major health risks to the local population [5]. The PMC was constantly forced to face the unsustainable nature of its disposal operations because of protests from impacted communities. Projects involving incineration were occasionally put forth as remedies, presented as "modern" and cutting edge. However, given the diversity of India's waste streams—which are low in calorific value and abundant in organic matter—such facilities were frequently financially unviable, capital-intensive, and opposed by locals and activists worried about air pollution.

It became more evident by the early 2000s that top-down, centralized techniques were unable to keep up with Pune's garbage realities. The collecting, transportation, and disposal phase of the linear paradigm was socially unfair, economically unsustainable, and environmentally damaging. Conversely, community-led and decentralized methods demonstrated promise. Pilot projects for neighbourhood composting showed that source segregation might greatly lower landfill loads [6, 7]. Large amounts of dry garbage were already being recovered and recycled by informal workers, but they were not given any credit or assistance for their efforts. The issue was not technical viability but political will: could decentralization be expanded to citywide operations and could the city incorporate informal labour into a formalized system?

With the mobilization of grassroots organizations, Pune's trash political economy started to change. In the 1990s, a waste pickers' trade union called the Kagad Kach Patra Kashtakari Panchayat (KKPKP) fought against their exclusion and called for acknowledgment of their economic and environmental contributions. The union emphasized how rubbish pickers supported their families through recycling revenue while also saving the municipality millions in avoided disposal expenses [8]. Their efforts reframed garbage as a social and political issue of labour, rights, and citizenship rather than just a technical one of disposal.

The SWaCH cooperative's seeds were planted in this contentious environment. The PMC was unable to invest reliably in large-scale incineration or extend landfills indefinitely, which had reached a breaking point for Pune's garbage economy. Conversely, informal pickers showed resilience and organizational ability [1, 9]. The groundwork for a new governance structure that would acknowledge waste pickers as valid service providers, decentralize processing and segregation duties, and lessen reliance on unsustainable centralised solutions was laid.

Before SWaCH, Pune's waste system was characterized by contradictions: a fast-expanding city producing mountains of rubbish, a shoddy municipal apparatus that couldn't handle the situation, and a sizable informal sector that provided crucial recycling services while going unacknowledged and marginalized [1]. Both catastrophe and opportunity were brought about by these tensions. They gave rise to an ingenious reaction that was based on collaboration, decentralization, and the conversion of social marginalization into environmental and civic leadership.

7.1.1 Life at the Margins: Waste, Work, and Exclusion in Pre-SWaCH Pune

Exclusion, environmental deterioration, and insecure livelihoods defined Pune's trash economy prior to the founding of the SWaCH cooperative. Solid waste volumes increased due to the city's rapid growth, while official municipal infrastructure fell far behind, resulting in chaotic disposal and incomplete collection. Thousands of informal rubbish pickers, often women from castes that have historically been marginalized, filled this void by establishing a parallel recycling system. They made a crucial but unacknowledged contribution while operating in hazardous, stigmatized, and unstable environments.

Waste pickers' demographics were a reflection of long-standing socioeconomic stratification trends. Many were members of oppressed castes, such as the Dalit, who have historically been forced to perform vile labour. In this workforce, women were particularly prevalent, juggling domestic duties with rubbish collection, and frequently serving as the main providers of income in urban poverty situations. Elderly family members and children were also enlisted to help pick trash from landfills, trash cans, and the streets. Waste work's gendered and caste-based aspects contributed to its invisibility by forcing what was considered "dirty" onto people who were already socially stigmatized, trapping them in cycles of exclusion [10–12].

The working environment was dangerous and harsh. Waste pickers frequently worked without gloves, boots, or masks, putting them at constant risk for infections, cuts from sharp items, and chronic musculoskeletal and respiratory issues. Hazardous chemicals, medical waste, and decomposing organic matter were among the concerns associated with handling mixed, unsegregated garbage [13, 14]. The meager and erratic incomes were reliant on the shifting rates scrap dealers paid for metal, glass, plastic, and paper. Only enough for subsistence could be obtained from a day's work, and there was hardly any bargaining power. Waste pickers held one of the most vulnerable jobs in the urban economy, with no official contracts, social protection, or recognition.

At the city's disposal sites, these working conditions developed against the backdrop of an environmental crisis. The Uruli Devachi landfill and other disposal sites were overflowing, badly designed, and not sufficiently inspected by the early 2000s. Leachate contaminated agricultural land and wells by leaking into groundwater. The air quality was dangerous due to open burning and methane emissions. Chronic health issues, ranging from skin conditions to respiratory disorders, were reported by the people living in neighbouring communities. Affected communities' frequent protests highlighted the landfill's hazards by portraying it as an environmental injustice and a public health disaster [1, 15].

Whole wards were underserved due to the municipality's, at best, sporadic collection mechanism. Maps from this era reveal significant coverage gaps, especially in low-income and outlying communities where garbage gathered in open sewers or dumps [16]. These collecting failures raised the risk of diseases like cholera and dengue and made flooding during the monsoon season worse. The only people frequently rerouting garbage away from these hazards were informal rubbish pickers, who worked outside of the municipal system. Despite taking the majority of the risks, they effectively became the city's unacknowledged environmental service providers.

These material hardships were exacerbated by social stigma. Waste pickers were frequently ejected from neighbourhoods, accused of stealing, or harassed by the police or locals. Their social mobility and access to housing, education, and finance were restricted by their affiliation with "dirty work", which served to uphold caste systems [1, 11]. The contradiction was glaring: those who kept Pune's recycling economy going and cut municipal expenses were denied the privileges of official citizenship.

Both ecological disintegration and human marginalization were characteristics of Pune's garbage system prior to the intervention. Municipal collection failed, informal workers performed necessary but subpar job, and landfills endangered the environment and public health. These circumstances paved the way for innovative ideas, but they also made it abundantly evident that any solution would need to address ingrained caste, gender, and class disparities in addition to trash volumes [16]. These circumstances led to the formation of the SWaCH cooperative, which turned marginalization into collective bargaining strength and invisibility into recognition.

7.2 Emergence of the SWaCH Cooperative and Zero-Garbage Pilots

A significant change in the conception, implementation, and politicization of waste management in an Indian metropolis was brought about by the establishment of the SWaCH cooperative in Pune. The city's sustainability goals reframed what had long been an unseen, marginalized task performed by men and women from excluded castes as a civic duty. The Kagad Kach Patra Kashtakari Panchayat (KKPKP) trade union, the PMC, and local communities worked together for decades to organize, negotiate, and experiment before this change was made [17]. The cooperative approach, which connected livelihoods, circularity, and decentralized urban governance in a way that went against accepted waste management paradigms, had not only received official recognition by 2007 but was also incorporated into policy.

The KKPKP union, formed in 1993, was the crucible for this change. The union organized rubbish pickers into a collective that could demand rights during a period when they were deeply stigmatized and exploited. The leadership of KKPKP, which was frequently derived from activist-academics and rubbish pickers themselves, changed the perception of pickers by framing them as environmental stewards and frontline recyclers rather than as a "problem." In addition to advocating for identity cards, credit availability, and welfare program participation, the union sought greater institutional recognition by persuading the municipality that its members were already saving the city a substantial amount of money by keeping tons of recyclables out of landfills each day [18]. Opening talks with the PMC, which had up until that point mostly disregarded or mistreated informal labourers, required this rephrasing.

Pressure on the PMC was increasing by the middle of the 2000s. While centralised incineration systems faced financial and legal issues, Uruli Devachi landfill crises had provoked public outrage [1]. Alternatives that were both economically viable and politically acceptable were required by municipal leaders. Seizing the opportunity, the union put forward a plan where waste pickers, formally contracted but organized under their own cooperative structure, would serve as the front-line providers of door-to-door collection and sorting services [14]. Because the PMC was hesitant to give a worker-led organization too much authority, negotiations were drawn out and frequently contentious [1]. However, the magnitude of the waste issue and the waste pickers' proven ability to organize gave them influence.

SWaCH (Solid trash Collection and Handling) Cooperative, a trash pickers' self-owned and self-managed organization, was born out of this convergence in 2007. The cooperative structure was a calculated move since, in contrast to NGOs or contracts, it positioned waste pickers as co-owners of their labour and services, with collective decision-making authority instead of going through outside parties. This represented a significant shift away from both strictly municipal provision and traditional privatization methods. By publicly acknowledging SWaCH as the partner in charge of primary collection, the PMC and SWaCH signed a memorandum of understanding [19]. Members of SWaCH were rewarded with legitimacy, training, ID cards, and uniforms.

At the same time, decentralized waste processing pilot projects started to appear in Pune's neighbourhoods. The Zero-Garbage Project, which was started in collaboration with NGOs and citizens' associations, sought to show that ward-level composting and household-level segregation could significantly lessen reliance on landfills. While SWaCH members directed dry garbage into recycling streams, organic waste was treated by nearby compost pits and biogas facilities. The cooperative's reasoning was supported by these experiments: decentralization was only successful if frontline employees possessed organizational competence and legitimacy [20]. When pilot initiatives were successful, they produced compelling evidence that recycling, composting, and segregation could cut trash to almost zero levels at the neighbourhood level.

Although it took some time for the cooperative to become institutionalized, it eventually became a part of city policy. By establishing user fees and distributing subsidies, PMC integrated the cooperative into its solid waste management system. The approach evolved into a hybrid public-community collaboration, with the cooperative mobilizing labour, expertise, and community networks while the municipality granted official approval and some financial support. This collaboration carried symbolic significance as well. Waste pickers were for the first time acknowledged as valid participants in international conferences, sustainability studies, and municipal policy documents rather than being marginalized.

The establishment of SWaCH also marked a significant advancement in resourceful urbanism: the integration of circular economy principles with social justice. Pune used existing informal systems and improved them through cooperative structures rather than making significant investments in costly incinerators or centralized mega-projects [21]. In contrast to many high-tech options, scarcity—in terms of technical capacity, landfill space, and municipal funds—was reframed as a chance to create a low-tech, inclusive, decentralized system that was more robust and adaptable.

Thus, the history of SWaCH's formation embodies a pragmatic and negotiating politics. Once unseen, waste pickers banded together to achieve recognition, the municipality, which had previously relied on malfunctioning landfill systems, looked for alternatives during a crisis, and they all formed a relationship that made it difficult to distinguish between the state and the community. By providing proof of concept, the Zero-Garbage pilots demonstrated that decentralized circularity was feasible rather than ideal. What started out as unofficial survival techniques was formalized as a cooperative that was incorporated into the city's system of government by 2007.

This move was not just about waste; it was also about redefining citizenship, redistributing power, and revaluing labour that had been undervalued for a long time. Through the SWaCH cooperative, marginalized workers were able to assert their right to economic stability, respect, and visibility while also providing Pune with an innovative and sustainable form of urban metabolism. It paved the way for the expansion of inclusive circularity in a rapidly urbanizing Indian city and symbolized the culmination of decades of ingenuity and sacrifice.

7.3 Cooperative Governance and Hybrid Finance: The Institutional Backbone of SWaCH

The unique governance and financial structure that underpins the SWaCH model are what give it its longevity, not just its technical advancements or social acceptability. SWaCH functions as a cooperative that is owned and run by waste pickers themselves, in contrast to contractor-based systems or fully municipal programs. This self-governing organization offers institutional legitimacy and autonomy as part of a public-community collaboration with the PMC [1, 22]. The hybrid system, however, is precarious since it must continually balance market volatility, state reliance, and the needs of maintaining services in a city that is expanding quickly.

Fundamentally, SWaCH operates as a cooperative organization. Many waste pickers who had previously worked informally joined as co-owners and members. They choose representatives, take part in decision-making, and influence laws that have an immediate impact on their quality of life. The cooperative structure was a conscious decision since it made sure that workers retained control of the organization rather than having it concentrated in outside NGOs or contractors. With features like elected committees, grievance redressal forums, and yearly general meetings, this system also promoted democratic accountability. Through the formalization of informal workers into a cooperative, SWaCH provided dignity and acknowledgment, integrating democratic practices into waste management operations on a daily basis.

Formal sanction was provided by the institutional relationship with the PMC. SWaCH was mandated to offer primary door-to-door rubbish collection when the city signed a memorandum of understanding with the cooperative in 2007. Under this arrangement, the PMC remained responsible for secondary transport and disposal, while the cooperative managed segregation, collection, and recycling at the household and neighbourhood level. A tiered system was institutionalized by this division of duties; the municipality-maintained supervision while acknowledging the cooperative as a valid participant in the provision of urban services.

This system's funding has always required careful consideration. The three main sources of funding for SWaCH are home user fees, municipal subsidies, and recyclables earnings. For door-to-door collection, households are encouraged to make small monthly contributions, establishing a clear connection between community accountability and service delivery. Even if they fluctuate, municipal subsidies aid in filling operating gaps and provide some financial stability for the cooperative [1]. Another source of revenue is recyclables, albeit this varies according to the unstable scrap markets. As a result, the cooperative maintains itself through a diverse portfolio—a sign of resourceful governance—rather than relying on a single source of revenue (see Table 7.1).

There are definite advantages to this hybrid financial approach. User fees introduce accountability: homes that pay SWaCH members directly are more willing to participate with segregation practices and develop a deeper sense of recognition for their labour. In addition to ensuring that services are not totally reliant on unstable markets, municipal subsidies lend legitimacy. Recycling revenue recovers value from waste streams that would otherwise be wasted, lowering landfill loads and providing support for livelihoods at the same time. When combined, these systems enable SWaCH to remain resilient and adaptable.

The system does have some weaknesses, though. The collection of user fees is frequently unstable. Particularly in low-income neighbourhoods where rubbish collection is supposed to be a free municipal service, many households put off or refuse to pay. The task of dealing with neighbours falls on waste pickers, which can sour relations and lower incomes. Despite being essential, municipal subsidies are usually delayed, making it difficult for the cooperative to cover its operating expenses. Uncertainty is also brought about by political changes within the PMC; adjustments in leadership or policy agendas may jeopardize contractual stability or interfere

Table 7.1 Institutional and financial architecture of the SWaCH cooperative—showing core mechanisms, enabling instruments, public and worker value, key fragilities, and how Pune has mitigated them in practice

Dimension	Core mechanism (what SWaCH does)	Instruments/examples	Value created	Fragilities/exposure	Mitigation in practice
Governance	Worker-owned cooperative delivers primary collection and segregation; PMC retains secondary transport/disposal	MoU with PMC (2007, renewals); elected committees; AGMs; grievance forums [1, 19, 22]	Legitimacy for waste pickers; service continuity; democratic accountability	Political turnover; contract uncertainty; elite capture risks	Regular PMC–SWaCH coordination; transparent minutes; leadership rotation rules; third-party facilitation via KKPKP [17, 22]
Finance	Blended revenue: household user fees + municipal subsidy + sale of recyclables	Ward-level fee schedules; subsidy line items; scrap auctions/longer-term buy-back	Cost recovery; income stabilisation; reduced municipal OPEX (transport/landfill)	Fee non-payment; subsidy delays; scrap price volatility	Linking fees to housing-society bylaws; escrow for subsidies; diversified buyers; emergency relief funds [1, 12, 18]
Operations	Door-to-door wet/dry segregation; decentralised composting/biogas; dry-waste aggregation and sale	Route plans; uniforms/IDs; neighbourhood compost pits and biogas; ward "zero-garbage" pilots	Higher recovery rates; fewer mixed loads; lower landfill intake and haul distances	Uneven compliance; equipment shortages; route creep	Behaviour campaigns; route rationalisation; basic PPE kits; society-level segregation mandates [14, 20, 24]
Social inclusion and labour	Formalising informal workers as co-owners and service providers	ID cards; trainings; welfare enrolment; women-led committees	Dignity, bargaining power, safer work; improved, more predictable incomes	Household conflict over fees; exposure to hazardous waste	Fee-collection protocols; joint society meetings; source-segregation enforcement; hazard reporting and PPE [11, 12]
Environmental outcomes	Material recovery + organics treatment at source; district-level circularity	Daily diversion of dry recyclables; neighbourhood compost use; methane avoidance	CO_2e reductions; cleaner streets; reduced leachate/odour at dumps	Performance uneven across wards; data gaps	Ward dashboards; independent audits; pilot-to-policy scaling with PMC support [1, 5, 20]
Scalability and transferability	Public–community partnership model vs. contractor-only or incineration-led	KKPKP union base; cooperative ownership; citywide MoU	Replicable principles (decentralisation, inclusion, blended finance)	Context dependence (municipal openness; organising capacity)	

The table summarises how SWaCH's cooperative governance, blended finance, and day-to-day operations convert marginalised labour into circular public value—along with the model's main fragilities and mitigations

with funding. SWaCH members are also subject to price volatility due to their reliance on recyclable markets, especially during periods of global commodity downturns.

Attempts have been made to reduce these risks through governance measures. The cooperative and the PMC communicate frequently in order to discuss delays and work out support. Additionally, it has created tools to improve household compliance, such as awareness campaigns and, in certain places, the connection of user fees to housing society agreements. The democratic governance systems in place within SWaCH are designed to preserve openness and guard against elite takeover. However, these systems need ongoing upkeep. There is a constant conflict between autonomy and reliance on local assistance: too much reliance runs the risk of co-optation, while too little dependency runs the risk of being left out of policy frameworks.

Notwithstanding these difficulties, SWaCH's financial and governance structure is a trailblazing example of equitable urban service delivery. It shows that informal labour can be institutionally integrated into formal governance systems through cooperatives, and that services can be sustained at scale through hybrid finance as opposed to only depending on markets or subsidies [23]. The model's longevity depends on its ability to preserve this equilibrium, guaranteeing that waste pickers continue to have authority, residents receive reliable service, and the municipality carries out its responsibility for environmentally friendly waste management.

Ultimately, the SWaCH cooperative serves as an example of how innovative urbanism may be institutionalized through the creation of inclusive, flexible, and community-owned financial and governance structures. Resilience is woven into the very fabric of the cooperative, and its success stems not from eradicating risk but rather from developing the ability to manage those risks collaboratively.

7.4 Turning Waste into Work: How SWaCH Delivers Circular Services in Practice

The everyday mechanisms of how a city can transform waste and exclusion into inclusion and value are made clear by the application of the SWaCH cooperative's concept. An arrangement between a rubbish pickers' cooperative and the PMC developed into a system that currently serves hundreds of thousands of households every day. Its power comes in fusing social recognition, community accountability, and low-tech but methodical operations to make circularity a lived practice rather than just an abstract idea.

Door-to-door collection is the foundation of SWaCH's activities. Currently employed by the cooperative, waste pickers are given predetermined routes that pass through individual homes, apartment buildings, and residential societies. In order to separate trash into "wet" (organic) and "dry" (recyclable) streams, they gather waste straight from the source, which is an essential innovation. In contrast to centralized methods that try to separate at transfer stations or landfills—typically with unsatisfactory outcomes—the SWaCH system starts at the household level, empowering collectors to enforce compliance while ingraining responsibility with residents. In addition to improving recovery rates, this change transforms the regular interactions between residents and staff into a mutually affirming environment [24].

The system includes decentralized processing and composting in addition to collection. Compost pits at the neighbourhood level, biogas plants, and "zero-garbage projects" convert organic waste near the place of generation into energy or fertilizer. These decentralized nodes strengthen local accountability for trash management while lowering transportation expenses and reliance on landfills. These experiments, which involve NGOs, ward-level authorities, and residents' associations, show how decentralization, when incorporated into collective governance, may scale in an urban setting, albeit unevenly.

Material recovery channels receive the dry waste that SWaCH members gather. An economic chain connects homes to local recycling markets through the sorting, aggregation, and sale of plastic, glass, paper, and metal to scrap dealers and recycling companies. What distinguishes SWaCH here is not the technology but the legitimacy of the workers who drive the system. Former informal pickers are now recognized as service providers rather than trespassers or scavengers thanks to uniforms, identity cards, and cooperative membership. As important as the actual material recovery is this symbolic change, which integrates visibility and dignity into the circularity infrastructure.

This system produces quantifiable and noteworthy results. Pune's landfill loads have decreased by hundreds of tons every day, and it's estimated that millions of kilograms are diverted annually. Decentralized composting facilities have shown significant decreases in organic waste, and recycling rates have increased. Benefits for employees are as significant: the cooperative offers collective bargaining strength, social safeguards are being expanded, and usage fees and recyclables stabilize incomes [1]. The city gains as well, demonstrating its leadership in inclusive circular economy principles and avoiding millions of dollars in transportation and landfill expenses.

Expensive technology or expansive infrastructure are not necessary for the successful implementation of SWaCH. Instead, it is found in the careful balancing of daily work, civic engagement, and local acknowledgment. Pune demonstrates that circular services may be provided efficiently and fairly by integrating resource recovery into neighbourhoods and acknowledging the labour of those who have previously been marginalized. The cooperative demonstrates that it is possible to completely restructure a city's metabolism, one neighbourhood and one door at a time.

7.4.1 Life on the Ground: Worker Experiences and Public Perceptions of SWaCH

The SWaCH cooperative's success cannot be determined solely by the amount of money made or the number of tons of waste that are diverted. The perspectives of the houses it serves and the lived experiences of its employees will be its true litmus test. Both the empowerment that comes with official acknowledgment and the conflicts that still exist at the intersection of labour, service delivery, and public expectations are revealed by day-to-day practice.

The cooperative has changed the way rubbish pickers do their everyday tasks. Participants attest that having official identity cards and uniforms gives them a sense of dignity that was lacking in previous years when they were frequently harassed for scavenging. In contrast to landfill picking's unpredictability, door-to-door collection methods offer more stable revenue streams through user fees and sales of separated recyclables [11, 24]. In contrast to former periods when they were made invisible, many employees observe that residents now address them by name, which is a little but significant change. But there are still dangers. It can be difficult to collect user fees since some homes refuse to pay or postpone payments, which puts workers at risk. Physical hazards also continue to exist since workers continue to come into contact with hazardous materials in household waste due to insufficient segregation, which emphasizes the necessity of constant resident compliance.

Though varied, residents' opinions are largely favourable. Many recognize the increased cleanliness and less roadside dumping, as well as the ease of doorstep collection. Housing societies have implemented wet-dry separation systems as a result of the cooperative's emphasis on segregation, which has also increased awareness. Nonetheless, there is a disparity in compliance, with middle- and upper-class households frequently collaborating better than low-income neighbourhoods where awareness and financial capacity are lower. Some locals are resentful of the little user fee since they believe that rubbish collection is a municipal duty that is already paid for by taxes. This conflict draws attention to a persistent difficulty in striking a balance between public acceptance and financial sustainability.

Because of SWaCH, Pune has frequently been hailed as a pioneer in urban sustainability in the media and in public discourse. The cooperative is cited by national and international media as an example of inclusive garbage management, and Pune is regularly mentioned at urban sustainability conferences as an example of how informal labour may be successfully integrated. The city is at the vanguard of discussions on circular economy techniques in the Global South, as evidenced by case studies that highlight the simultaneous accomplishment of environmental benefits and worker empowerment.

Critiques, however, moderate this story. Depending on local political will, community collaboration, and municipal assistance, performance is still inconsistent among wards. A change in municipal leadership or a delay in funding could destabilize the cooperative, making it extremely vulnerable to legislative changes.

Despite advancements, some activists contend that waste pickers are still paid too little for the vital nature of their labour and that stronger institutional guarantees are necessary for long-term sustainability.

In sum, daily living in SWaCH is characterized by both enduring precarity and hard-won dignity as well as apparent service improvements [25]. Although the cooperative's worth is acknowledged by the public, acceptance is still reliant on local cooperation and political backing. These facts, which remind us that circularity is not only technical but also profoundly social and is constantly negotiated between employees, residents, and the government, anchor the cooperative in actual reality.

7.5 Beyond Waste: Evaluating Impact and Weighing the Limits of the SWaCH Model

The reason the SWaCH cooperative has gained notice is because it produces quantifiable outcomes in an industry that is frequently beset by inefficiency and invisibility. The cooperative has changed Pune's waste metabolism on a quantitative level. With about 3800 members, SWaCH provides daily doorstep collection services to millions of citizens, serving over half of the city's households. According to estimates, hundreds of thousands of tonnes of rubbish are diverted from landfills annually, with hundreds of thousands of tonnes removed daily. According to independent calculations, the cooperative saves more than 100,000 tonnes of carbon dioxide a year, and municipal reports attribute this to the cooperative's assistance in preventing landfill disasters [1]. These numbers highlight the importance of a worker-led, decentralized model for the environment.

The qualitative effects are just as potent as the quantitative ones. Cooperative participation has resulted in a significant identity change for waste pickers, who have gone from being marginalized "rag pickers" to becoming acknowledged service providers and environmental stewards. Cooperative ownership, ID cards, and uniforms have all contributed to social recognition and dignity. These workers are now seen as genuine service providers by households who previously disregarded or stigmatized them. Waste pickers now have more access to social protections and more negotiating power in local politics as a result of this. Additionally, municipal capacity has benefited: the Pune Municipal Corporation has achieved greater recycling and composting rates than the majority of Indian cities while avoiding the prohibitive expenditures of large-scale landfill expansion or incineration projects by cooperating with SWaCH.

However, consideration of externalities is necessary for critical evaluation. Waste pickers, who continue to face physical hazards due to inadequate segregation and irregular user fee collection, are a key component of the cooperative concept. Each ward has a different level of service quality, with citizens' compliance varying and local council members' support being inconsistent. Workers now face financial precarity as a result of activities being occasionally disrupted by the municipality's payment delays. The approach is nonetheless susceptible to political changes at the structural level; for example, a new waste management contractor or a change in municipal leadership could jeopardize the cooperative's mission. The fine line between exposure and empowerment is highlighted by this reliance on a precarious contract.

On the other hand, SWaCH's longevity shows resilience. Since its establishment in 2007, the cooperative has endured changes in city leadership, municipal politics, and recycling market swings. Its longevity indicates that waste pickers' institutionalization inside a cooperative framework has produced an organizational structure that is flexible enough to adjust to shifting conditions. However, permanency is not assured by longevity. Continued public support for segregation and user fee compliance, increased financial stability, and strengthened municipal commitment are all necessary to maintain the paradigm.

In terms of transferability, SWaCH provides important lessons for Global South cities facing similar challenges of rapid urbanisation, rising waste volumes, and entrenched informality. Decentralization and social inclusion are two entwined ideals that are essential to its success. Both labour rights and service delivery are addressed when informal workers are given ownership of the system and integrated into formal government. Decentralized recycling and composting lessen reliance on landfills and show that efficient waste management doesn't have to be expensive. Transferability is not always the case, though. Strong community involvement, a comparatively open municipal partner, and decades of union organizing (via KKPKP) are prerequisites for the SWaCH model that might not be present everywhere. Instead of copying the model exactly, cities that want to imitate Pune must modify it to fit their own institutional, political, and social circumstances.

In conclusion, SWaCH is a practical example of resourceful urbanism; it shows how a city may prosper despite social disparity, landfill saturation, and financial constraints by enlisting the help of the community, forming institutional alliances, and implementing decentralized systems. The cooperative's symbolic successes in redefining garbage work as respectable, acknowledged labour are equal to its quantifiable effects, which include waste diversion, CO_2 savings, and improvements in livelihood. Its durability gives us cautious hope, and its externalities serve as a reminder that resourcefulness is not without costs. Pune's SWaCH experiment serves as both an inspiration and a challenge as cities around the world struggle with the twin crises of inequality and waste. It serves as evidence that inclusive, decentralized systems can function and a reminder that ongoing negotiations between the state, labour, and the community are essential to their survival.

References

1. Malhotra M (2001) Social impact assessment of the Pune waste collection model: the Swach movement. Doctoral dissertation, Symbiosis International University
2. Saifi N, Jha B (2024) An overview of solid waste management practices in Pune, Maharashtra, India. Nat Environ Pollut Technol 23(2):923–934

3. Estrada M, Galvin M, Maassen A, Hörschelmann K (2023) Catalysing urban transformation through women's empowerment in cooperative waste management: the SWaCH initiative in Pune, India. Loc Environ 28(7):852–866
4. Mastakar P, Mastakar V, Mukhopadhyay K, Jaju S (2019) Zero waste circular model of MSWM: a success in Ward 40, Pune, India. Metamorphosis 18(1):36–56
5. Perlman RMK (2015) Identifying sustainable organic management systems in urban India: case study of Pune, India. Doctoral dissertation, Massachusetts Institute of Technology
6. Chikarmane P, Vidyavihar MK (2014, April) Of users, providers and the state: solid waste management in Pune India. In: putting public in public services: research, action and equity in the global south. Municipal services conference, Cape Town, April, pp 13–16
7. Bhat V, Patil Y (2021, July) An integrated and sustainable model for e-waste management for Pune City households. In: Journal of physics: conference series (vol 1964, no. 6). IOP Publishing, p 062111
8. Butsch C, Kumar S, Wagner PD, Kroll M, Kantakumar LN, Bharucha E, Schneider K, Kraas F (2017) Growing 'smart'? Urbanization processes in the Pune urban agglomeration. Sustainability 9(12):2335
9. Rao KC, Parthan B, Doshi K (2018) Power from municipal solid waste at Pune municipal corporation. Pune
10. Gadgil R, Menon S (2018) Solid waste management and women's livelihoods in Pune: a desired future. Ahmedabad: Center for Environmental Education. 48 pages report. Available from https://the3dprogram.org/content/uploads/2018/10/3D-Program-SWM-Report-September-2018.pdf
11. Cheshmehzangi A (2025) Waste and material management: optimizing resource flows for Advanced City's sustainability agenda. In: Advanced City: for the future of cities and opposing smart cities. Springer Nature Singapore, Singapore, pp 455–465
12. Chikarmane P (2012) Integrating waste pickers into municipal solid waste management in Pune, India. WIEGO Policy Brief (Urban Policies) 8:23
13. Failor T (2010) Improving services and improving lives: waste-picker integration and municipal coproduction in Pune, India. Available from https://cdr.lib.unc.edu/concern/masters_papers/02870x421
14. Dawodu A, Oladejo J, Tsiga Z, Kanengoni T, Cheshmehzangi A (2022) Underutilization of waste as a resource: bottom-up approach to waste management and its energy implications in Lagos, Nigeria. Intell Build Int 14(3):261–282
15. Sohkhlet D, Nagargoje S (2020) Municipal solid waste management: a comparative study between Sydney (Australia) and Pune (India). In: E3S Web Conf, vol 170, p 04001. EDP sciences. Available from https://www.e3s-conferences.org/articles/e3sconf/abs/2020/30/e3sconf_evf2020_04001/e3sconf_evf2020_04001.html
16. Shankar VK, Sahni R (2020) Integrating the informal with the formal: a case of cohesive development in urban waste chains. In: Theorizing cohesive development. Routledge, pp 174–185
17. Chikarmane P, Narayanan L (2023) Pune waste pickers' innovative efforts to fund the extension of decent work and social protection. Resource Document, WIEGO
18. Parhi S, Siom P (2018) Solid waste management: a study of Pune municipal corporation. Pune, SIOM, 1–17
19. Radhakrishnan M, Pathak TM, Irvine K, Pathirana A (2017) Scoping for the operation of agile urban adaptation for secondary cities of the global south: possibilities in Pune, India. Water 9(12):939
20. Ghosh P, Bobba ER, Dodda SS, Jasti VK, Meka S, Vanga HR (2025) Scenario of solid waste management in Indian Cities: a study of Pune, Visakhapatnam, and Tirupati. Ecol Econ Soc–the INSEE J 8(1):55–76
21. Cheshmehzangi A, You N, Siri J, Birch E (2024) Promoting sustainable environment conservation and women empowerment through Faecal sludge and Septage management (FSSM) in Berhampur, India. In: Harnessing urban innovation to unlock the sustainable development goals. Springer Nature Singapore, pp 67–72, Singapore

22. Mohan G, Sinha UK, Lal M (2016) Managing of solid waste through public private partnership model. Procedia Environ Sci 35:158–168
23. Ritter WF, Dsouza C (2014) Social and environmental impacts of waste management alternatives in India: a critical analysis. In: World environmental and water resources congress 2014, pp 1588–1597
24. Sakipour S (2011) Evaluation of opportunities in e-waste processing facility investment in Pune, India. Available from https://www.theseus.fi/bitstream/handle/10024/37411/Sakipour_Sara.pdf;sequence=2
25. Sundar KS (2019) Globalization, labour market institutions, processes and policies in India. Palgrave Macmillan, Singapore

Part IV
Multifunctional Urban Spaces

Chapter 8
Landscapes of Multiplicity: Yokohama's Zoorasia and Forest Park Network

8.1 From Sprawl to Stewardship: Context and Political Economy of Yokohama's Multifunctional Landscapes

The story of Yokohama's Zoorasia and Forest Park Network cannot be told without first tracing the city's post-war transformation and the pressures it produced on land, ecology, and public life. The second-biggest city in Japan, Yokohama, expanded quickly as a port and an industrial center following World War II. Its peripheries were being reshaped by suburbanization and metropolitan growth by the 1960s and 1970s, when housing estates, factories, and arterial roads had replaced farms and forests [1]. Once home to satoyama, the traditional managed forests that supported village economies and wildlife, the peri-urban borders were now degraded and fragmented [2]. Rapid expansion, which fuelled economic prosperity, also put a strain on natural systems and reduced the amount of public green space that a growing population could use.

New needs were created by this rapidly urbanizing environment. By the 1980s, local communities and authorities realized that unregulated development was endangering both natural health and human well-being [2]. Leisure and environmental quality are becoming major urban concerns due to rising wealth and educational attainment. While politicians recognized the need to limit flood hazards, restore biodiversity, and lessen urban heat, residents wanted access to recreational grounds. In contrast to central Tokyo, Yokohama's peripheries included land that might be transformed into multipurpose landscapes [3–7], but doing so would need long-term planning frameworks and political commitment.

The resuscitation of satoyama was essential to this vision. Traditional mosaics of grasslands, rice paddies, and woodlands, these environments have long sustained local livelihoods and biodiversity [8]. However, as labour transitioned away from land-based economies due to industrialization and urban migration, many satoyama had been abandoned by the end of the twentieth century [9]. As a result of their

A. Cheshmehzangi, *Resourceful Urbanism*, Urban Sustainability,
https://doi.org/10.1007/978-981-95-4138-6_8

collapse, habitats were lost, but a cultural practice that struck a balance between natural renewal and human usage also vanished. In the 1990s, as part of a larger sustainability agenda, national and local policy frameworks started to advocate for satoyama restoration. Yokohama was in a good position to incorporate this policy change into urban planning because of its distinctive combination of peri-urban land and civic aspirations [10].

This convergence of opportunities and demands was reflected in the choice to invest in Zoorasia and the Forest Park that surrounds it. The proposal was presented by the city of Yokohama as a landscape system that would provide ecological, social, and infrastructure advantages, rather than just a zoological park. Planning plans from the late 1990s characterize the zoo as a node in a larger forest network rather than as a stand-alone attraction [1, 2]. The adjacent Forest Park would improve water regulation by integrating with floodplains, restoring satoyama habitats, and creating pathways for bicyclists and pedestrians [11, 12]. According to this perspective, peri-urban green space was a type of green infrastructure that was necessary for community health, biodiversity, hydrology, and cooling.

Politically, the project complemented Japan's changing urban sustainability strategy. The relevance of "eco-cities" and environmental education was emphasized in national initiatives during the 1990s, and the local administration of Yokohama aimed to establish itself as a pioneer in these areas. The city was already known for its large-scale urban initiatives, such the renovation of Minato Mirai 21 [13, 14], and Zoorasia/Forest Park gave it the opportunity to show its leadership in environmental conservation. A key player in furthering this goal was the Yokohama Greenery Foundation, an institutional entity created to oversee and promote the city's green resources. The project was positioned as an example of multifunctional urbanism in action [15] by combining recreation, education, and conservation.

Additionally, the larger political economy was involved. As Japan struggled with the fallout from its economic bubble in the 1990s, communities looked for investments that would provide long-term benefit without resorting to the excesses of speculative urbanism. When compared to large-scale infrastructure projects, peri-urban green systems were comparatively inexpensive, but they yielded observable and quantifiable benefits in terms of ecosystem services, public health, and tourism. In particular, Zoorasia was created to connect environmental objectives with cultural and economic development to draw millions of tourists there each year. A strong political rationale for the project was made possible by the combination of ecological restoration and tourism income [16].

In summary, Yokohama's Zoorasia and Forest Park Network was born at the nexus of growing social needs for green infrastructure, ecological deterioration, and suburban encroachment. Peri-urban land has been purposefully reframed as a multipurpose system supporting recreation, education, biodiversity, and climate resilience, rather than as underutilized or degraded land [1, 2]. The project's political economy depended on acknowledging that green spaces are infrastructures, not extravagances, that can provide various types of public benefit in a time of ecological crisis and financial restraint. In addition to allowing the project to proceed, this conceptualization guaranteed its applicability to broader discussions about how cities may transform fringe settings into resources that support ecosystems and people.

8.1.1 *Landscapes in Limbo: The Peri-Urban Edge Before Zoorasia*

The northern outskirts of Yokohama were neither entirely rural nor entirely urban prior to the establishment of Zoorasia and the Forest Park Network [1]. They were landscapes in a state of flux, torn between the steadfast expansion of suburban sprawl and the ruins of traditional satoyama management. This "in-between" state is crucial to comprehending the significance of the undertaking. It was constructed on disjointed terrain that ran the risk of falling into neglect rather than on a virgin forest or in the middle of an urban area.

This fragmentation was reflected in the land usage prior to the intervention. Secondary forest patches and agricultural areas coexisted, both of which displayed symptoms of abandonment as rural livelihoods deteriorated. While some parcels were overgrown, others were used as makeshift parking lots or dumping sites. It was neither barren nor fertile—neither heavily cultivated nor environmentally sound. Rather, it had the signs of disconnection and underuse that were common in many late twentieth-century Japanese peri-urban areas. Isolated tree stands, asymmetrical fields, and growing road networks pushing into the terrain are all visible in the patchwork shown in satellite photos taken at the period.

Concern over ecological deterioration was growing. Once supported by cycles of grazing, coppicing, and agricultural integration, Satoyama had been neglected for decades. In the absence of proactive management, native biodiversity decreased, alien species spread, and forest health deteriorated [9]. Conventional mosaic landscapes were disintegrating into uniform thickets after supporting a high species richness. This area was particularly affected by habitat loss, which was already a national problem as fields and woodlands were engulfed by urbanization. Roads and suburban subdivisions disrupted wildlife routes, making fragmented habitat pockets more susceptible to additional disturbance.

The growth of suburbs exacerbated these environmental stresses. As Yokohama's population increased in the 1970s and 1980s, there was a need for transportation, housing, and schools [17, 18]. Subdivisions and road extensions were absorbed by the northern boundary, which turned into a development border. The residual woodlands now serve more as leftover buffers than as cohesive green systems, fragmenting what could have been a continuous biological corridor. Due to its close vicinity to expanding neighbourhoods, the site was constantly under pressure to either be developed or abandoned.

At the same time, there were noticeable gaps in public access to green spaces. Despite being known for its center redevelopments and waterfront parks, Yokohama's northern neighborhoods lacked significant, easily accessible leisure space [1]. There were few options for communal gatherings in natural settings, outdoor learning, or walking for locals. When there were parks, they were frequently little, well-kept, and isolated from the wider biological area. There were few easily available opportunities for families and schools to learn about biodiversity or environmental education in their local area. The perception of peri-urban land as marginal rather than as a resource in and of itself was strengthened by the lack of significant green infrastructure.

This feeling of liminality is captured in photographs and early planning documents. Not abandoned enough to require immediate action, yet not alive enough to satisfy the ecological and social demands of the city, the area that is now a renowned zoo and forest park system was once a landscape of compromise. In essence, it was an area that needed to be rethought.

This liminal condition was also an opportunity. The potential for rehabilitation was left open by the lack of heavy development. Despite their degradation, the scattered satoyama landscapes retained their cultural resonance and ecological memory [19, 20]. It would be possible to resurrect traditional knowledge of woodland management and agricultural integration as a useful foundation for multipurpose landscapes in a contemporary metropolis, rather than as nostalgia. Because of its peri-urban location, the area was both open and roomy, making it suitable for both major recreational initiatives like Zoorasia and more modest ecological interventions like corridor construction and forest restoration.

In hindsight, the pre-intervention circumstances highlight the goal of the initiative. Often written off as peripheral, the peri-urban edge was reframed as a key area where urban life, culture, and nature might converge. Instead of erasing a flourishing landscape, Zoorasia and the Forest Park Network stepped in to bring it back into harmony while adding additional ecological, social, and educational uses [1, 2]. This transformation's tale is as much about restoring neglected landscapes as it is about building new infrastructure.

8.2 Crafting a Living Landscape: The Genesis of Zoorasia and the Forest Park Vision

The development of Zoorasia and the nearby Forest Park Network in Yokohama was the result of careful planning, changing urban priorities, and an understanding that the city's future required innovative approaches to fusing ecological and culture. Yokohama was already regarded as one of Japan's most forward-thinking cities by the 1990s, thanks to significant revitalization initiatives like Minato Mirai 21 [13]. In addition to its sparkling shoreline, the city's peri-urban areas were beset by ecological deterioration, fragmented green spaces, and a dearth of easily accessible public landscapes in its rising northern districts. To overcome these obstacles, more was needed than just another city park. It called for a rethinking of green space as a living infrastructure capable of serving biodiversity, education, recreation, and climate adaptation simultaneously.

The City of Yokohama's planning documents from the 1990s reveal this shift in priorities. Urban officials realized that providing green space could not be restricted to well-kept, centrally located parks. Growing worries about urban heat, biodiversity loss, and climate change necessitated the creation of multipurpose landscapes that might serve as ecological resilience corridors, classrooms, and lungs [1]. The city simultaneously aimed to develop its leisure and cultural offerings in ways that would attract tourists and benefit locals' education. One possible anchor was a large new zoo, designed as a conservation-focused enterprise rather than a show. Planners envisioned a project that would blur the lines between urban infrastructure and ecological restoration, recreation and education, by placing the zoo inside a larger forested setting.

This concept was developed in large part by the Yokohama Greenery Foundation, which was founded to protect and promote the city's natural resources. The foundation developed a design philosophy that aimed to preserve satoyama woodlands while adding new uses, working closely with local authorities [8, 21, 22]. Zoo planners and landscape architects were hired to create an interconnected network where cultural programs and biological processes supported one another, rather than just creating exhibits or walkways [1, 2]. From ecologically sensitive zoo enclosures to restored woodland patches, from outdoor learning spaces to walking trails, from water management systems to kid-friendly play areas, the focus was on establishing continuity.

The hybridity of the design concept was ambitious. Neither Zoorasia nor Forest Park was designed to be a stand-alone zoological facility or a passive green buffer. Rather, the initiative aimed to integrate them into a cohesive system, with cultural and recreational programming serving as a layer on top of ecological restoration, which served as the basis. The zoo was designed to serve as an educational entryway, drawing guests into an experience that focused more on relating to larger ecosystems and conservation stories than it did on watching animals in cages. Its proximity to the Forest Park allowed guests to expand their experience into volunteer-led forest care projects, outdoor schools, and walking and cycling routes.

A characteristically Japanese sensibility toward landscapes was represented in this integration: a recognition of satoyama as cultural-ecological legacy and an understanding of the connectivity between the natural and built surroundings [23]. Instead of being presented as sentimental, the restoration of these historically managed woodlands was presented as a practical solution to the stresses of urban climate change and biodiversity loss. Early on in the planning phase, volunteers and citizen organizations were included in the hope that continued community involvement would be necessary for long-term stewardship. Through the integration of cultural initiatives like outdoor education, festivals, and citizen science into the landscape, the project made sure that ecological restoration had a social foundation.

Another rationalisation came from tourism and education. It was intended that Zoorasia would bring 1.5 to 2 million guests a year, including both domestic families and visitors from all across Japan. Its modern, conservation-focused zoo design set it apart from other zoos in the country, and its affiliation with the Forest Park guaranteed that the tourism industry would profit from a broader network of recreational and ecological opportunities [1, 2]. Since the location serves as a living classroom for environmental education, schools were also recognized as important stakeholders.

Finally, Yokohama's desire to show leadership in multifunctional urbanization was mirrored in the design concept for Zoorasia and the Forest Park. It was a hybrid of a traditional park and a traditional zoo, built on ecological restoration and layered with recreational and cultural elements. Through its dual placement as a gateway and a destination, Zoorasia set an example for the reclamation and reimagining of peri-urban territory, not as unused space but as essential urban infrastructure that supported biodiversity, bolstered community identity, and taught future generations.

8.3 Governing the Green Mosaic: Institutions and Finance Behind Zoorasia and the Forest Park

A large-scale multifunctional landscape may be maintained when civic actors, semi-independent foundations, and public leadership work together toward common objectives, as demonstrated by the financing and governance of Yokohama's Zoorasia and Forest Park Network. Unlike ad hoc parks or commercially driven attractions, Zoorasia and the surrounding forest network were intended to be long-term civic assets. As such, they needed governance structures that could preserve financial viability while striking a balance between ecological purity and visitor participation.

At the centre of this architecture is the City of Yokohama. The zoo and the surrounding parklands were established by the municipality, which served as the lead authority and supplied the original plan, land distribution, and capital expenditure [24]. The project was presented by city officials as an infrastructure investment in health, environment, and education rather than as a standalone recreational amenity. Overall supervision, long-term planning integration, and making sure the project complies with Yokohama's cultural and environmental regulations are still the responsibility of municipal departments.

The Yokohama Greenery Foundation, a semi-public organization charged with overseeing and promoting the city's natural resources, supports this municipal leadership. The foundation plays a crucial role in mediating between conservation objectives and recreational usage, developing volunteer and instructional programs, and providing day-to-day management experience. Because of its semi-independent position, it can form alliances with private donors, community organizations, and NGOs while still being answerable to the municipality. The foundation is able to experiment with programs while keeping them rooted in public goals thanks to its dual standing between official authority and citizen involvement.

The project has been financed using a meticulously layered model. The city's dedication to ecological infrastructure was demonstrated by the fact that the first phases were financed by public investment. The system expanded its sources of income when it was up and running. Zoorasia's ticket sales generate a consistent flow of revenue that supports the zoo's operating expenses as well as more general ecological and educational initiatives within the park system. Municipal budgetary allotments, which support ecological restoration, infrastructural improvements, and long-term upkeep, supplement this. NGOs, educational partners, and a vibrant volunteer base provide further support, as their voluntary labour lowers expenses while enhancing civic ownership. In this way, the financial model captures the multifaceted, multilayered, and adaptable nature of the terrain itself.

The governance model has to strike a balance all the time. On the one hand, ticket sales are crucial for Zoorasia's financial sustainability because the park welcomes over a million visitors a year. However, an over-reliance on tourism could jeopardize biodiversity and weaken the ecological goal of the project. The major tools for preserving this balance are visitor management techniques, such as restricting access to delicate habitats, dividing flows among several zones, and including educational materials into leisure activities [1]. Visitors are guaranteed to be both leisure consumers and participants in conservation awareness thanks to the Greenery Foundation's educational outreach.

But dangers still exist. Large zoo and park systems have enormous operating costs, and as the COVID-19 pandemic clearly showed, relying too heavily on tourism income leaves the project open to economic downturns or public health emergencies. Typhoons, heat waves, and habitat stress are examples of climate vulnerabilities that present long-term issues that could raise maintenance costs and call for additional ecological adaptation. These dangers emphasize the necessity of sound financial planning, which may involve more varied sources of income or finances for climate adaptation.

The financial and governance framework supporting Zoorasia and the Forest Park has held up well in spite of these difficulties [2]. Yokohama has developed an institutional architecture that is both accountable and flexible by combining community involvement, cooperative administration via the Greenery Foundation, and state funding. It proves that multifunctional green infrastructure needs governance models that can manage trade-offs, maintain cash flows, and strike a balance between ecological objectives and political and economic realities.

8.4 Weaving Ecology and Experience: Implementation and Outcomes of Zoorasia's Landscape Network

The integration of public programming and ecological restoration into a unified, multipurpose system is exemplified by Zoorasia and the nearby Forest Park Network. In order to ensure that biodiversity, recreation, and education were mutually reinforcing, Yokohama's planners and designers created continuity between the zoo, park, and neighbouring woodlands rather than considering them as distinct entities.

The project's integration of zoo displays with the forest park pathways is one of its distinguishing features. In order to minimize the feeling of artificial isolation, Zoorasia's enclosures were constructed to mimic natural habitats and fit in with the surrounding woodland [1]. Visitors can easily transition from animal exhibits into restored satoyama landscapes thanks to pathways that stretch outside the zoo into the Forest Park. This physical integration marks a conceptual change: the zoo is a doorway into larger ecological systems rather than a spectacle cut off from nature. Table 8.1 summarises the network's multifunctionality, i.e., what is built, how value is measured, and how risks are managed, so the outcomes that follow can be read against concrete levers rather than abstractions.

One of the main interventions has been the restoration of satoyama habitats [25]. Through thinning, replanting, and community-led management, traditional managed forests that had previously been neglected were brought back to life. Native plants and animals found homes in restored forest, grassland, and wetland mosaics, which restored biodiversity. Important resources in a metropolis dealing with urban heat and climatic stresses, these interventions also enhanced biological processes like soil retention, flood control, and microclimate cooling.

Designing for people was equally important. The environment featured pathways for bicyclists and walkers that connected outdoor classrooms, entertainment spaces, and zoo displays. Playgrounds and picnic areas were available to families, and educational facilities incorporated into natural environments were available to students. A school trip, a weekend family outing, or a volunteer forest-care activity all took place within the same multipurpose framework, creating a system where leisure and education were indistinguishable.

Recreation is just one aspect of the programmatic results. While visitor surveys demonstrate heightened awareness of environmental issues, biodiversity indicators provide quantifiable benefits from habitat restoration. There is a high level of community involvement, with volunteers participating in citizen science and forest management. There are also observable health and climate co-benefits: shaded corridors promote exercise, absorb stormwater, and lower local heat. In short, execution created a multi-layered municipal asset out of fragmented peri-urban terrain. The Forest Park and Zoorasia were constructed as living infrastructures—areas where human well-being and ecological resilience are co-produced—rather than as monuments.

Table 8.1 Multifunctionality Map of Yokohama's Zoorasia–Forest Park Network—i.e., How the network delivers ecological, hydrological, educational, recreational, economic, and governance value—paired with example indicators, key risks, and the levers used to keep trade-offs in check

Function strand	Concrete interventions	Example indicators/how to measure	Primary beneficiaries	Key risks/trade-offs	Management levers
Biodiversity and habitat restoration	Satoyama thinning and coppice renewal; native understory replanting; wetland/paddy mosaics; habitat-mimetic zoo enclosures linked to forest edges	Native species richness; % native ground/understory cover; habitat connectivity index; invasive species incidence	Urban wildlife; schools/researchers; citywide ecosystem health	Trampling/edge effects from high footfall; invasive rebound without ongoing care	Seasonal trail closures; volunteer habitat crews; invasive control cycles; buffer plantings
Hydrology and climate regulation	Floodplain reconnection; riparian planting; shaded trail corridors; permeable paths	Peak flow attenuation (storm events); soil moisture retention; surface temperature delta on heat days	Downstream neighbourhoods; park users during heatwaves	Extreme rain/typhoon damage; rising maintenance costs	Green–grey drainage upgrades; emergency repair funds; climate-risk design standards
Education and cultural literacy	Outdoor classrooms; citizen science; interpretive signage; satoyama craft days	Annual learner-hours; # classes hosted; volunteer hours; survey scores on eco-literacy	Students/teachers; families; civic groups	Program fatigue; quality drift if budgets tighten	Co-delivery with Greenery Foundation/ NGOs; docent training; curriculum tie-ins
Recreation and public health	Continuous walking/cycling network; play lawns; picnic zones; quiet loops	Trail km accessible; use counts; user satisfaction; proximity (15–20-min access)	Local residents; visitors; seniors' groups	Overcrowding; path erosion; user conflicts (cyclists/pedestrians)	Zoning by intensity; wayfinding; path hardening where needed; timed events
Tourism and local economy	Zoorasia admissions; seasonal festivals; farm-stand/market tie-ins	Annual visitors; dwell time; local spend; cost-recovery ratio	City finances; small vendors; hospitality	Overtourism pressures on habitat; sensitivity to shocks (pandemics)	Visitor caps in sensitive zones; diversified revenue (memberships, grants)
Governance and finance	City leadership + Yokohama Greenery Foundation operations; mixed revenue (tickets + municipal budget + donations/volunteers)	Operating coverage (%); capital upkeep backlog; volunteer contribution (FTE-equiv.)	Whole system (keeps lights on)	Budget volatility; mission drift toward entertainment	Medium-term funding compacts; ring-fenced ecology line items; annual mission audit

8.4.1 Living with the Landscape: Public Encounters and Everyday Reception

The greatest way to gauge the effectiveness of Yokohama's Zoorasia and Forest Park Network is to look at how people use, appreciate, and experience the area on a daily basis rather than just biodiversity stats or design prizes [1]. Millions of tourists interact with the landscape in ways that support its multipurpose function, allowing the project to become ingrained in the rhythms of urban life.

The number of visitors highlights its importance. Zoorasia is one of the most visited zoological sites in Japan, with between 1.5 and 2 million visitors annually. Together with the nearby Forest Park trails and parks, the network serves as a significant public resource that attracts tourists, school groups, and families. High levels of satisfaction are found in surveys by the City of Yokohama and independent researchers, who highlight the special fusion of education, recreation, and natural immersion. For many locals, the area provides a unique opportunity to leave the crowded suburban neighbourhoods and enter a setting that feels both local and global—a setting steeped in satoyama customs but linked to worldwide conservation issues.

Passive visitation is not the only usage in daily life. A hallmark of the Forest Park's operations is volunteerism, with neighbourhood organizations working on citizen science initiatives, habitat monitoring, and forest care. Outdoor classrooms are a common feature of school excursions, where students can learn about cultural heritage, biodiversity, and climate change firsthand. For these groups, the park serves as a living laboratory as well as a green area. As much of an effect as the actual restoration of habitats is the civic culture of engagement.

Critiques also highlight the conflicts that arise in multifunctional environments. Although the location is well-connected by road, some groups, particularly the elderly and those without private transportation, find it difficult to be included due to the site's dependency on car travel. Another constant problem is striking a balance between conservation and the number of visitors. High yearly foot traffic strains delicate ecosystems and erodes walkways, which worries conservationists that if popularity is not controlled, it may jeopardize ecological objectives.

The Forest Park and Zoorasia have received overwhelmingly positive reviews in spite of these conflicts. For the people of Yokohama, it is now more than just a park or zoo; it is a communal civic space that blends recreation, education, and stewardship. Its daily use shows that when biological systems are conceived to be both habitat and habitat-for-people, multifunctionality may be successful.

8.5 Beyond the Zoo: Impacts, Trade-Offs, and Lessons from Yokohama's Living Landscape

The Zoorasia and Forest Park Network has developed into one of Yokohama's most recognizable civic landscapes 20 years after it was first created. It has a noticeable effect on quantitative, qualitative, and symbolic dimensions, but it also brings up significant issues regarding expenses, environmental stresses, and the long-term viability of big, multipurpose undertakings. Analyzing its results provides a sobering acknowledgement of the constraints and compromises inherent with resourceful urbanization, as well as a celebration of what has been accomplished.

The numbers are outstanding from a quantitative standpoint. Between 1.5 and 2 million people visit Zoorasia each year, and thousands more come to enjoy the nearby Forest Park network's walking, cycling, and recreational opportunities [1]. These amounts of visitors attest to the project's status as a significant cultural and recreational destination, on par with the waterfront revitalization of Yokohama. While forest thinning and wetland restoration have improved water regulation and decreased flood hazards in nearby suburban districts, biodiversity monitoring indicates a rise in native species inside restored satoyama ecosystems. These ecological indicators lend credibility to the narrative. Additionally, according to microclimate research, the green lanes help to significantly lower local urban heat. These measurements offer concrete proof that the landscape serves as infrastructure in addition to being a recreational area.

The qualitative effects are equally significant. Through citizen science, forest stewardship, and volunteer programs, the initiative has promoted new kinds of community cohesion. The restoration of Satoyama has given residents a greater feeling of place and cultural identity, tying urban life to a more profound ecological legacy. As environmental knowledge is incorporated into regular schooling through field trips and outdoor classrooms, ecological literacy has grown. A distinct sense of pride is revealed by public surveys: Zoorasia is more than just a zoo; it is a representation of Yokohama's dedication to sustainability and cultural innovation.

However, there will always be externalities in a project this size. Sensitive ecosystems experience ecological stress due to high visitor numbers, which forces management to continuously modify access and maintenance schedules. Another issue is financial sustainability: although Zoorasia's ticket sales mostly pay for operating expenses, the larger park system still depends on volunteer labour and municipal funding. Cost recovery and long-term resilience in periods of financial restriction are called into doubt by this reliance on public funds and unpaid involvement. The peri-urban setting also presents challenges: biological corridors are still in danger from adjacent development, and the site's dependence on automobile access restricts its inclusivity and raises its carbon impact.

The project's lifespan is remarkable despite these drawbacks. The network has withstood changing social norms, economic downturns, and political environments since it first launched in the late 1990s. Resilience is demonstrated by its flexibility in incorporating new educational initiatives, increasing volunteer opportunities, and responding to climate threats. Where many similar projects fail, the governance mechanism, which is anchored by the Yokohama Greenery Foundation and municipal monitoring, has made sure that things continue.

From a global standpoint, the lessons are obvious. Resourceful urbanization can benefit greatly from multipurpose landscapes like Zoorasia and the Forest Park, which demonstrate that peri-urban land need not be viewed as speculative or left-over land but rather can be reframed as ecological and social infrastructure. Education, enjoyment, and restoration work together to provide synergies that are more effective than the sum of their individual components. However, transferability necessitates prudence. Japan's strong traditions of civic engagement, local government, and cultural observance of satoyama landscapes were essential to Yokohama's prosperity. Instead of trying to replicate these ideas, cities overseas must modify them to fit their unique political, cultural, and ecological circumstances.

In summary, Zoorasia and the Forest Park Network serve as an example of an ingenious urban intervention, showing how deteriorated peri-urban borders can be restored to create liveable landscapes that strike a balance between leisure, culture, and ecology. Their effects go beyond statistics; they also change how people view the environment, their neighbourhoods, and the identity of their cities. The trade-offs—costs, environmental impact, and accessibility issues—remind us of that multifunctionality is never without its obstacles. However, the long-term success of Yokohama's strategy shows that cities can create landscapes that benefit people and the environment for many generations when they treat green spaces as infrastructure rather than decoration.

References

1. Japan Travel (2019) Guide to visiting Zoorasia: one of Japan's best zoos, Community writer: Cathy Cawood. Available from https://en.japantravel.com/kanagawa/zoorasia/58924
2. Yamazaki K, Dollery B, Kinoshita Y (2024) Local factors sustaining co-production: two case studies from the city of Yokohama, Japan. J Urban Aff 46(5):1039–1052
3. Hein C (2010) Shaping Tokyo: land development and planning practice in the early modern Japanese metropolis. J Urban Hist 36(4):447–484
4. Sato H (2025) Planning a 'sustainable city' with hinterlands: the case of Yokohama City, Japan and its soft space planning to reach net-zero. Environ Plann E Nat Space 8(2):721–741
5. Hao W (2023) Tokyo, global city of cinema 1980–1995: cultural infrastructure and the urban imaginaries of globalisation. Doctoral dissertation, University of Warwick; Nagoya Daigaku
6. Edgington DW (1994) Information systems in Japanese Cities and Regions. Planning for Cities and Regions in Japan, 1, 126
7. Hein C (2014) Port cities and urban wealth: between global networks and local transformations. Int J Glob Environ Issues 13(2–4):339–361
8. Takeuchi, K., Brown, R. D., Washitani, I., Tsunekawa, A., & Yokohari, M. (Eds.). (2008). Satoyama: the traditional rural landscape of Japan. Springer Science & Business Media
9. Ito T, Sugiura M (2021) Satoyama landscapes as ecological mosaics of biodiversity: local knowledge, environmental education, and the future of Japan's rural areas. Environ Sci Policy Sustain Dev 63(5):14–25

10. Cheshmehzangi A, You N, Siri J, Birch EL (2024) Harnessing urban innovation to unlock the sustainable development goals. Springer, Singapore
11. Ariga T (2025) Japanese environmental design and Management for Sustainable Urbanism: learning from Satoyama. Taylor & Francis, Abingdon
12. Adesina JA, Tang X, Uduma-Olugu N, Adebamowo M (2025) Regenerating urban landscapes through wetlands restoration and outdoor open space connectivity: off-road bicycle lane planning principles for tourism development. Turismo: Visão e Ação 27:e20521
13. Suzuki T (2013) Downtown revitalization through public private partnerships: the case of Minato-Mirai 21, Yokohama, Japan. Master's thesis, Tufts University
14. Balaban O (2013) The use of indicators to assess urban regeneration performance for climate-friendly urban development: the case of Yokohama Minato Mirai 21. In: Spatial planning and sustainable development: approaches for achieving sustainable urban form in Asian cities. Springer Netherlands, Dordrecht, pp 91–115
15. Cheshmehzangi A (2025) [Urban] planning and pathways to better management in advanced cities. In: Advanced City: for the future of cities and opposing smart cities. Springer Nature Singapore, Singapore, pp 75–85
16. Murakami K (2025, April) Policy approaches to sustainable and circular economy tourism: comparative case studies of Japan and China. In: Proceedings of the 7th international conference on tourism research. Academic Conferences and publishing limited
17. Nakamura A (1996) The "Tokyo problem" and the development of urban issues in Japan. Jun/Wright (1996) 191–203
18. Sorensen A (2001) Subcentres and satellite cities: Tokyo's 20th century experience of planned polycentrism. Int Plan Stud 6(1):9–32
19. Ishizawa M (2018) Cultural landscapes link to nature: learning from Satoyama and Satoumi. Built Herit 2(4):7–19
20. Takeuchi K, Ichikawa K, Elmqvist T (2016) Satoyama landscape as social–ecological system: historical changes and future perspective. Curr Opin Environ Sustain 19:30–39
21. Takeuchi K (2003) The nature of Satoyama landscapes. In: Satoyama: the traditional rural landscape of Japan. Springer, Tokyo, pp 9–39
22. Terada T, Yokohari M, Bolthouse J, Tanaka N (2010) Refueling Satoyama woodland restoration in Japan: enhancing restoration practice and experiences through woodfuel utilization. Nat Cult 5(3):251–276
23. Dublin DR (2015) "Satoyama-Satoumi" regional management: a universal cognition and practice for green economy, ecosystem health and sustainable society in indigenous communities of the world. Doctoral dissertation, 北海道大学, Hokkaido University
24. Taguchi T (2022) The post-war rebirth of Yokohama: the planner Akira Tamura's contributions to municipal reform. Plan Perspect 37(5):1073–1095
25. Takeuchi K (2013) Rebuilding satoyama landscapes and human–nature relationships. Sustain Sci Series 244

Chapter 9
Resourceful Social Urbanism: Medellín's Library Parks as Engines of Collective Transformation

9.1 From Narco-City to Knowledge City: Political Economy of Medellín's Library Parks

It is impossible to comprehend the emergence of Medellín's Library Parks without first placing them in the context of the city's most troubled decades. Medellín gained international notoriety in the 1980s and 1990s as the home of Pablo Escobar and the Medellín Cartel. Systemic violence resulted in homicide rates that rose to some of the highest in the world, with over 350 murders per 100,000 inhabitants in some years. Daily life was tinged with fear, and entire neighbourhoods—especially the barrios erected on the hillside by themselves—became associated with narco-territorial domination [1]. The violence was exacerbated by stigma; Medellín was not just viewed as violent but also as unredeemable.

Inequality and exclusion were severe at the same time. Some people became wealthy as a result of the city's fast industrialization in the middle of the twentieth century, while others were forced to live in outlying communities that were perched on steep slopes. These barrios were politically and geographically marginal because they lacked safe public areas, schools, and healthcare, and they had inadequate access to the city center [2]. In addition to housing instability, informality meant that the state was not present. The only institutions that many locals could see were those of the black market.

During this time, municipal capacity was severely limited. Corruption was rampant, government was disjointed, and budgets were tight. The supply of infrastructure reinforced spatial inequality by concentrating on core regions and highways. Traditional urbanism, such as massive transportation corridors, large-scale engineering, or technocratic planning, provided few solutions for communities in which the most pressing needs were opportunity, dignity, and trust rather than efficiency.

A new political initiative started to emerge in this setting. By the early 2000s, Medellín was experimenting with what became known as "social urbanism", led by

A. Cheshmehzangi, *Resourceful Urbanism*, Urban Sustainability,
https://doi.org/10.1007/978-981-95-4138-6_9

Mayor Sergio Fajardo and a group of reform-minded individuals. The idea was surprisingly straightforward: construct infrastructure for social inclusion and cultural change in addition to efficiency and mobility [1, 3]. In actuality, this meant transforming marginality into opportunity by focusing public funds on the neighbourhoods most affected by neglect and violence.

This change was symbolized in the Library Parks. The city combined parks and libraries into multipurpose infrastructures rather than viewing them as specialized educational spaces or recreational areas. With its combination of library collections, classrooms, digital access, plazas, gardens, and cultural events, each Library Park was intended to serve as a colossal anchor in a hillside neighbourhood. Architecture had a major role; prominent Colombian architects created bold, recognizable designs that were prominently displayed in the city's most vulnerable areas rather than concealed in affluent neighbourhoods [4]. The message was clear: just as much as people in wealthy neighbourhoods, the people living in these barrios should have access to top-notch public facilities.

The inventiveness at work is brought to light by seeing this through the prism of resourceful urbanism. The municipal authority of Medellín lacked the financial resources to construct redundant facilities for public space, education, culture, and recreation. Rather, it aimed to consolidate several tasks into a single infrastructure component. A Library Park served as a school, a plaza, a safe haven, a computer hub, and a cultural center all at once. The public value per peso invested was maximized by this tiered design, which is a crucial strategy in a time of financial constraint [5]. Additionally, it made sure that residents engaged with the institution in a variety of ways, such as by participating in community events, playing in the plaza, taking a literacy lesson, or borrowing a book.

Thus, the Library Parks' political economics was a rearranging of urban priorities rather than just architectural aspirations. The city moved resources away from technical modernization and toward social rehabilitation by integrating investment in marginal areas. It claimed that institutions that promote knowledge, belonging, and civic pride may create public value in addition to roads and bridges by presenting culture and education as infrastructure.

The Library Parks came to represent Medellín's turnaround story on a global scale. The Library Parks represented the city's resolve to reconstruct itself through inclusivity, in addition to the cable cars that connected hillside barrios to the city center and the urban escalators that sewed mobility into difficult slopes [5, 6]. They were both political statements and useful interventions at the same time; they represented a new Medellín that was characterized by creativity, culture, and group potential rather than violence.

A deeper lesson can be learned from the history of the Library Parks when viewed through the prism of resourceful urbanization. Importing high-tech solutions or copying international "best practices" won't bring about change in environments of violence and shortage. It results from rephrasing limitations into design briefs: marginality as a location for investment rather than abandonment, fragmented governance as a catalyst to involve communities, and restricted resources as a drive to layer functions [6]. Medellín's Library Parks were about more than just books and structures; they were about recovering the city's right via inclusive, multipurpose, and creative urban planning.

9.1.1 Before the Libraries: Marginal Hillsides, Violence, and the Absence of Institutions

Medellín's hillside barrios, which were eventually selected for the development of Library Parks, were paradoxical places: full of community networks and human energy, but also characterized by fear, exclusion, and a lack of even the most basic municipal amenities. It is crucial to first outline these pre-intervention circumstances to comprehend the Library Parks' transformative importance.

Hundreds of thousands of people lived on the slopes of Medellín by the end of the twentieth century, many of them displaced families from Colombia's rural conflicts. In self-organized communities called comunas, people erected their houses on steep slopes with minimal formal planning [7, 8]. These homes were usually improvised, consisting of corrugated metal roofs and walls made of brick and cinderblock, frequently perched perilously on landslide-prone terrain. There was little infrastructure for secure electrical connections, water supply, or drainage, and the streets were small, often consisting solely of meandering dirt roads. There was very little public infrastructure, including few clinics, packed schools, and almost no parks. Children played in alleys or on rooftops instead than in parks or plazas.

These barrios' restricted mobility added to their geographical isolation. It could take more than an hour to take the costly and unreliable bus services from the steep hillside communities to the city center. The geography of exclusion was glaring: the barrios remained socially and physically isolated from Medellín's center, which benefited from investments in roads and trade. The state seemed to be absent save for occasional police raids, which furthered the feeling of desolation.

These neighbourhoods were characterized by significant structural exclusion at the same period. Many kids dropped out of school early because they couldn't afford it or didn't want to help support their families. There were few job prospects, particularly for young people, who were constantly tempted and under pressure from the illegal economies that flourished in the absence of authority [5]. Due to the widespread absence of public institutions, such as libraries, schools, and cultural centers, there was both material and metaphorical deprivation. Implicit but obvious was the city's belief that these citizens did not deserve the same facilities and opportunities as other citizens.

Other types of authority filled this institutional gap. The hillside barrios were frequently the sites of fighting in the 1980s and 1990s, when Medellín came to be associated with narco-violence. From Escobar's cartel to later paramilitary groups, armed groups claimed control of territory [5, 9]. For the locals, this meant that dread and violence were commonplace. Daily routines were influenced by gang shootouts, targeted murders, and the ongoing danger of compulsion. Youngsters were used to hearing gunfire at night. Parents were afraid to let their kids play outside. It is impossible to overestimate the psychological toll of this environment: trauma, mistrust, and resignation became ingrained in society.

Ironically, informal administration by cartels or armed players occasionally offered the sole appearance of "order" even as violence characterized these areas. Sometimes local fiestas, sports teams, or basic welfare were sponsored by drug lords [10]. However, this increased the city's stigma toward the barrios and came at the expense of deference to their authority. These neighbourhoods were seen as dangerous by outsiders, but to the locals, they were just home. They survived by supporting one another and being resilient, but they lacked the resources and credibility of official municipal organizations.

These circumstances produced a contradictory environment in which violence and shortage both hampered progress and increased the need for immediate action. The very lack of a secure public area demonstrated the transformational power of a plaza, cultural center, or library. The potential of educational infrastructure as a lifeline was highlighted by the dearth of easily accessible schools. Given the barrios' remote location, it was possible that a well-thought-out municipal anchor could serve as a hub for both services and symbolic city integration.

To put it another way, limitations turned into the design brief. The city was compelled to think of multipurpose areas that could simultaneously address civic identity, education, culture, and recreation due to a shortage of infrastructure. Due of limited mobility, facilities had to be integrated within the neighbourhoods rather than located outside of them. Because there were no institutions, architecture had to be more than just structures; it had to be a representation of presence, dignity, and belonging. Public space was reframed by violence itself: a secure plaza outside a library was a weapon against fear rather than merely a recreational area.

Therefore, the pre-intervention circumstances in the hillside barrios of Medellín served as catalysts rather than merely backgrounds [11, 12]. Their fear, alienation, and hardship exposed the shortcomings of traditional urbanism and provided an ideal environment for an innovative strategy. Medellín sent a message that these areas should not be abandoned by transforming marginal hillsides into locations for audacious municipal development [13–16]. They would serve as the platform for the city to show that ingenuity, which combines public space, culture, and education into a single infrastructure, could transform not only the built environment but also the story of a whole city.

9.2 From Stigma to Symbol: The Birth of Medellín's Library Parks and Their Design Vision

A city devastated by violence and neglect ventured to redefine its most marginalized neighbourhoods as places of dignity, knowledge, and civic pride when Medellín's Library Parks emerged in the early 2000s. The concept emerged as part of a larger political culture change in Medellín, spearheaded by reformist politicians who were committed to ending the cycles of marginalization [5]. Mayor Sergio Fajardo (2004–2007), a mathematician who entered politics on the platform of "Medellín, the most educated", was at the center of this change [17, 18]. Under his leadership, the city's problems were reframed as chances for innovation through what is now known as resourceful urbanism rather than as insurmountable obstacles.

The foundation of Fajardo's vision was a straightforward but bold idea: provide the best public infrastructure to the most underserved communities. The typical development strategy, which had long prioritized investment in affluent neighbourhoods while ignoring the hillside barrios, was purposefully reversed in this case [5]. Education and culture were not ancillary services to Fajardo and his group; rather, they were vital components of the infrastructure needed to restore the social fabric. Accordingly, the Library Parks were intended to be more than just structures; they were also intended to serve as representations of civic potential, social inclusion, and state presence.

The design philosophy that underpinned the programme reflected resourcefulness in multiple dimensions. The municipal budget of Medellín was constrained, therefore there was no room for several distinct facilities with specific uses. Rather, multifunctionality—one structure, multiple lives—was the answer. Every Library Park was intended to function as a public plaza, a cultural center, a library, and a classroom all at once. The city created venues that could accommodate a variety of requirements by layering functions to maximize the social worth of every peso spent. In the morning, a youngster might go to a reading program; in the afternoon, a teenager could use the internet; and in the evening, a family could congregate in the plaza. The Library Parks were anchored as live municipal institutions by this density of use, which also brought about efficiency and vibrancy.

This goal was both practical and symbolic, and architecture played a crucial role. The administration of Medellín deliberately chose to hire famous, imposing designs from top Colombian architects. Projects like the España Library Park, designed by Giancarlo Mazzanti, with its striking black rock-like volumes perched dramatically above Santo Domingo Savio, exemplified this philosophy. These were not modest community centres tucked away on side streets; they were bold statements, visible landmarks intended to project dignity onto communities that had long been dismissed as dangerous or disposable. The city used architecture to assert that the underprivileged slopes should have buildings on par with those found in any major city in the world.

The Library Parks were positioned strategically as well. They were integrated into the neighbourhoods most impacted by exclusion and violence, turning them from places of stigma into areas of focus for investment. However, the vision went beyond the actual structures. The Metrocable gondola system and, eventually, hillside escalators were two of Medellín's other innovative infrastructure projects that were intended to be integrated with the Library Parks. The city increased the impact of its efforts by connecting new modes of mobility with cultural and educational facilities [5]. In addition to having easy access to local cultural and educational possibilities through a Library Park, residents could now effortlessly commute from a remote hillside barrio to the city center. Medellín's concept was characterized by this synergy between social infrastructure and physical connectedness.

Importantly, the Library Parks encompassed programming in addition to physical space. In order to guarantee that the facilities featured civic events, digital training, arts workshops, and literacy programs, Fajardo's administration collaborated with cultural organizations, educators, and community organizations. The legitimacy of the parks depended on this programming; otherwise, the buildings would have been meaningless symbols that locals would have ignored but visitors would have adored. The Library Parks transformed how communities interacted with information and public space by integrating cultural and educational events into daily life.

Thus, political will, creative design, and symbolic architecture came together to create the Library Parks. They began politically with a pledge to make investments where the city had previously made withdrawals. They embodied multifunctionality in their design, making the most of few resources [18]. In terms of architecture, they brought dignity to areas that had previously only been characterized by neglect and terror. Additionally, they were included strategically into larger mobility enhancements, guaranteeing that social infrastructure was interwoven with the city's fabric rather than being in a vacuum.

From the perspective of resourceful urbanism, the Library Parks transformed Medellín's biggest limitations into architectural possibilities. Budget constraints forced multifunctionality. Integration with transportation systems was necessary due to geographic isolation. Bold architectural statements capable of rewriting narratives were necessary to combat stigma. The necessity of secure, welcoming, and respectable public areas was highlighted by violence and marginalization. In addition to a collection of structures, a new paradigm for how cities may use education and culture as a means of repairing society was born.

Medellín's Library Parks shown that the most creative urban interventions are not those that spend the most money, but rather those that dare to turn the least resourced areas into the richest sources of civic value by transforming barrios that were formerly defined by terror into places of pride.

9.3 Building Trust in Brick and Balance Sheets: Governance and Finance of Medellín's Library Parks

Although Medellín's Library Parks are frequently praised for their eye-catching architecture and symbolic location in underprivileged barrios, their continued viability as active public organizations depends as much on governance architecture and funding as on design. These initiatives were woven into a complicated web of institutional collaborations, community involvement, and municipal leadership rather than existing as standalone monuments. The inventiveness and vulnerability of Medellín's experiment with resourceful urbanism are demonstrated by the methods in which resources were mobilized and responsibilities were distributed.

The program's launch was spearheaded by the municipality of Medellín. The city took a risk by allocating limited public monies to the Library Parks under Mayor Sergio Fajardo. This was not a city with plenty of money to spare; Medellín had limited tax income and urgent needs for housing, transportation, and security. However, by deliberately allocating funding for educational and cultural infrastructure, the administration reframed these expenditures as investments in long-term societal stability. This institutional change emphasized the resourceful urbanism philosophy of achieving more with less by treating culture and education as "hard" infrastructure, as necessary as roads or bridges.

Leveraging relationships outside of city hall was necessary to finance the Library Parks. In order to guarantee that the libraries were more than just structures but were bustling hubs of learning, education agencies helped with program design and operating expenses. Medellín's daring rebranding as a social innovation laboratory drew international philanthropists, development banks, and cultural organizations. A portion of the training, digital infrastructure, and programming was funded by the Inter-American Development Bank and charitable foundations. Although they served as catalytic supplements and helped stretch municipal budgets and diversify revenue sources, these outside contributions were not the main source of funding.

The function of cultural institutions and community organizations was equally significant. In addition to serving as a library, each Library Park served as a center for civic programming, including literacy classes, art workshops, and public gatherings, many of which were organized in collaboration with neighbourhood NGOs and community organizations. With the municipality providing the facility and core finance and communities and cultural actors enlivening the space with programs, this distributed governance model established a type of cooperative guardianship. For legitimacy, this balance was essential. Without the support of the municipality, the parks ran the risk of being underfunded local hubs, and without community input, they ran the risk of being perceived as top-down mandates.

As a result, the governance structure was a triangle that connected institutional knowledge, community voice, and municipal authority. Although decision-making processes differed from park to park, they frequently involved advisory boards or ways for community members to participate and influence programming. Accountability was strengthened through community engagements, surveys, and annual general meetings. This interactive dimension helped ground the Library Parks in the everyday lives of locals rather than the lofty goals of planners, setting them apart from other civic initiatives.

However, there were hazards associated with both finance and governance. The biggest obstacle was probably political turnover. The continuity of funding became unstable since Fajardo's successors did not always share his aims. Budget cuts for programming at some Library Parks resulted in inconsistent network performance. The reliance on charismatic leadership highlighted a structural weakness: resourceful urbanism may rely too much on human champions in the absence of institutionalized safeguards.

Funding gaps also posed challenges. Construction was funded by municipal investment, but continuing operations necessitated steady funding for personnel, upkeep, and programming. Although small revenue-generating activities and ticketed events were helpful, cultural institutions cannot solely rely on business rationale. Although useful, the dependence on outside funders produced uncertainty when priorities changed or subsidies ran out. Parks occasionally found it difficult to maintain the level of programming that had previously made them successful.

Finding a balance between sustainability and scale was another concern. Despite receiving international praise, the imposing architecture required more upkeep. The focus on iconic form ran the risk of taking money away from less obvious but no less important services, as several critics pointed out. Others were concerned that, despite their symbolic strength, the parks would overestimate the potential of cultural infrastructure in the face of systemic injustices.

However, the governance and finance model's ingenuity was exposed right here. Even in the face of financial strain, the Library Parks showed resiliency by integrating programming into community partnerships, layering donor and municipal contributions, and diversifying financing sources. The actual currency that kept the system afloat was trust—between the government, the populace, and cultural organizations.

The Library Parks demonstrate that resourceful urbanization involves both institutional inventiveness and astute design in the larger context of Medellín's change. They demonstrate how participation can be incorporated into governance and how financing can be expanded by redefining investment as a group rather than an individual endeavour. In its prime, the parks transformed each peso into a variety of social, cultural, and educational benefits. When they were at their most vulnerable, they demonstrated how brittle these trials are when there is political unpredictability and financial constraints. In the end, Medellín's Library Parks' management and funding demonstrate that establishing faith in brick, balance sheets, and community relationships is just as important to urban development as creating libraries.

9.4 Architecture of Hope: Implementation, Design, and Programmatic Outcomes

Through a meticulously planned process that combined architectural ambition with programmatic layering, the Library Parks of Medellín went from concept to reality. Their implementation tells the tale of how everyday programming and constructed form were creatively blended to turn marginalized barrios into places of belonging and hope, in addition to the story of construction.

The architectural design was at the center of this change. With its striking shape visible from the hillsides, each Library Park was intended to be a beacon. Architects such as Giancarlo Mazzanti signalled strength and durability with stunning geometries, such as the España Library Park's black stone-like blocks. These were not small neighbourhood amenities; rather, they were imposing constructions that were purposefully meant to contradict the story of neglect. However, their grandeur went beyond mere beauty. With interiors that house adaptable areas for classrooms, libraries, digital access points, and multipurpose halls, the buildings were intended to be functional. Here, architecture took on two roles: usefulness and symbolism, physical dignity.

The Library Parks' location was just as cleverly planned. Many were situated where informal housing had once clung tenuously to steep, unstable terrain. Open plazas were carved into these slopes by designers, who did not view the topography as a constraint. These squares developed into level, secure areas where neighbours might congregate, kids could play, and public gatherings could take place. These common areas developed into new stages of communal life in places where fear had previously deserted the streets. They also addressed the entitlement to a public place that is both safe and accessible, which is a fundamental urban inequity.

The architecture was animated via programming. The Library Parks have never been meant to be only book warehouses. Daily routines included civic education, digital training, literacy sessions, and art workshops. The spaces developed into centers for cultural expression, adult education, and after-school activities. Alongside regular library activities, there was a place for community gatherings, theatre rehearsals, and musical performances [5]. The parks maximized their influence by combining different activities under one roof, resulting in multipurpose areas that addressed civic, cultural, and educational needs simultaneously. Table 9.1 summarises how each hillside constraint is translated into built form, programming, and measurable outcomes—plus the levers Medellín used to keep the model resilient.

The parks most obviously demonstrate resourceful urbanization in this layering. Due to financial constraints, Medellín was unable to make independent expenditures in community centers, schools, and plazas on each hillside. Rather, the Library Parks were intended to serve as a one-stop shop for cultural enrichment, public space, educational infrastructure, and civic pride symbols. The scope of one building's multi-sector participation much exceeded what its budget might indicate.

The results of this design approach were observable. Because of the well-lit plazas and the presence of cultural institutions that deterred illegal behaviour, residents said they felt safer. Children now have access to instructional materials that were previously unavailable in their communities [5]. Festivals and other activities that strengthened a sense of local identity were held by communities. Together, the public space, programming, and architecture created settings that not only met practical requirements but also changed how locals interacted with their city.

Implementation, in this sense, involved creating new social realities as well as physical structures. The Library Parks showed how cultural infrastructure can serve as a civic commons, an education system, and a method to reduce violence when it is thoughtfully planned and executed. The sharpest slopes in the city were transformed from places of exclusion into places of promise for everyone.

Table 9.1 Library Parks Logic Model: From Constraint to Collective Value - How Medellín's Library Parks convert hillside constraints into layered design, programming, measurable outcomes, and risk controls

Hillside constraint	Resourceful design move	Built components	Program layers	Example indicators	Key risks and the lever to manage them
Violence and fear in public space	Make dignity visible; anchor safety with activity	Iconic library volumes; lit plazas; sightlines; mixed-use edges	Evening cultural events; youth sports/ arts; community watch partnerships	Plaza footfall (day/night); perceived safety in resident surveys; reported incidents within 400 m	Risk: night-time underuse. Lever: evening programming calendar + targeted lighting upgrades
Educational exclusion	Stack functions in one civic hub	Classrooms; reading rooms; multiuse halls	Literacy, tutoring, test prep; maker/ media labs	Library card sign-ups; learner-hours; school completion proximate to parks	Risk: program drift. Lever: annual curriculum MoUs with education dept.
Digital divide	Put connectivity where stigma once lived	Public PC labs; Wi-Fi mesh; device-lending	Digital skills; job search; coding/media workshops	Unique users on Wi-Fi; certifications earned; job placements	Risk: equipment downtime. Lever: SLA with tech partner + spares budget
Public-space deficit on steep slopes	Sculpt flat, shared ground	Benched terraces; universal-access ramps; weather canopies	Markets, festivals, seniors' mornings, kids' play	Dwell time; event counts; age/gender balance of users	Risk: erosion/ maintenance. Lever: slope stabilization fund + seasonal closures
Mobility isolation	Tie social infra to mass transit	Park siting at Metro cable/ feeder links; bike parking	Wayfinding to transit; off-peak youth passes	Trips to parks via cable; first/ last-mile counts	Risk: affordability. Lever: fare concessions tied to park ID
Stigma and weak civic identity	Use form as narrative repair	Landmark architecture; neighbourhood-facing facades	Local art residencies; oral-history, community media	Participation in cultural programs; media sentiment analysis	Risk: icon without inclusion. Lever: quotas for local co-production

(continued)

Table 9.1 (continued)

Hillside constraint	Resourceful design move	Built components	Program layers	Example indicators	Key risks and the lever to manage them
Fiscal scarcity	One building, many lives	Flexible halls; divisible rooms; shared back-of-house	Day–evening turnover; fee-supported events	Space utilization rate; cost/benefit per user	Risk: budget shocks. Lever: blended finance (city + donors + earned)
Institutional fragility	Distributed stewardship	Municipal lead + library system + NGOs + community councils	Co-governance boards; volunteer corps	Volunteer hours; program continuity across administrations	Risk: political turnover. Lever: 3–5-year funding compacts; ring-fenced O&M line

9.4.1 *Living the Library: Everyday Experiences and Public Reception*

The Library Parks were more than just skyline landmarks for the people who lived in Medellín's hillside barrios; they were an integral part of their everyday lives, neighbourhood rhythms, and changing civic identities. Their reaction illustrates how physical structures may change how people live, see, and tell stories about their own communities when combined with well-considered programming.

For many locals, the opening of a Library Park marked the first time their neighbourhood had a piece of world-class infrastructure. The recurrent subject of *dignidad*—a sense of dignity restored—is captured in interviews by local researchers [19, 20]. Parents talked about how proud they were to walk their kids into a structure that people travelled from all over the city or possibly the world to see. Adolescents talked about the previously unthinkable access to computers, music classes, and study areas. The Library Parks turned into a daily source of inspiration.

These impressions were supported by survey data. The number of people signing up for library cards, attending community seminars, and participating in arts programs has increased, according to municipal evaluations [5, 18]. The use of public spaces increased as well; plazas surrounding the Library Parks saw an increase in foot traffic and were thought to be safer than previously. It was a quiet revolution to move toward active use of public space in neighbourhoods that had long been ruled by fear.

The Library Parks brought Medellín to the attention of the world. Featured in award ceremonies, movies, and urbanist conferences, they came to represent the city's "social urbanism" [18]. Pictures of their brash shapes, particularly the España Library Park, went around as proof that a city was taking back its destiny. The fact that the world was finally viewing them as communities deserving of respect and investment rather than as violent enclaves strengthened the sense of belonging for many locals.

However, not every perception was unquestioning. Some locals drew attention to accessibility issues, pointing out that some neighbourhoods were still neglected and that elderly and disabled users frequently had difficulty with steep approaches. Others pointed out that the enormous scope occasionally conflicted with the little resources available for maintenance, resulting in problems like malfunctioning machinery or a shortage of personnel. Following governmental changes, certain parks saw inconsistent programming, which raised concerns about sustainability that went beyond the news.

Nevertheless, the Library Parks' ongoing significance is evident in day-to-day living [21]. They provided a daily interaction with dignity, opportunity, and recognition—more than just books or plazas. Despite their flaws, the Library Parks continue to serve as potent reminders of how creative urbanism is experienced [2, 5, 22, 23] by regular people stepping through their doors rather than in policy documents.

9.5 Beyond Books: Impacts, Constraints, and Transferable Lessons

The Medellín Library Parks are among the most well-known instances of how cities may transform times of scarcity into chances for revitalization. Their influence has been quantified, recounted by locals, and honoured internationally. However, their story also provides crucial warnings on political continuity, sustainability, and the perils of symbols overtaking substance. They shed light on the potential and constraints of considering culture and education as infrastructure, serving as a chapter in the larger story of resourceful urbanism.

Their praise is well supported by quantitative indicators. Measurable decreases in violence were observed in neighbourhoods where Library Parks were implemented, according to municipal data. Although causality is complicated, connections indicate that more youth programs and safer public areas helped reduce homicide rates in areas that were previously associated with insecurity. Additionally, school enrolment increased in barrios with Library Parks, bolstered by literacy programs, tutoring, and digital tool access. With thousands of new members being added annually and cultural activities attracting locals who had not had access to the library before, library use itself has significantly increased. Tens of thousands of residents were served by these facilities each year, changing the way municipal services were provided throughout the city.

The qualitative results were possibly more significant than the quantitative ones. People frequently talked about feeling empowered and that the city finally recognized their worth as worthy of funding. As neighbourhoods started to be associated with cultural programs, safe plazas, and iconic architecture rather than terror, civic identity was reinforced. Statistics cannot capture the pride of showing off "our library" to guests or of seeing one's neighbourhood featured on postcards and in foreign media [5]. The city's social fabric underwent a psychological reweaving, giving once marginalized residents a new sense of identity.

However, recognizing externalities and limits is necessary for critical appraisal. Particularly following shifting municipal priorities due to political leadership changes, several parks experienced unequal programming or maintenance. Despite its strength, symbolic architecture has drawbacks. If large structures were not adequately funded, they might go underutilized and became impressive shells rather than thriving hubs. There were still disparities in accessibility, with certain populations remaining geographically remote or unable to fully engage because of financial, linguistic, or mobility constraints. Investing in some places had the risk of igniting gentrification pressures, which begs the issue of who stood to gain from the renewed focus on the city.

Both community ownership and institutional resiliency have been essential to the Library Parks' existence. The network's survival for almost 20 years despite political and economic ups and downs is evidence of how much the public cares for it. Because of their active use, residents have prevented these institutions from fading into obscurity. However, maintaining the Library Parks' full potential necessitates ongoing reinvestment, not only in physical assets but also in relationships, people, and programs. Here, resourceful urbanization is a continuous process of renewal rather than a singular accomplishment.

The lessons are both encouraging and a warning to other communities. First, Medellín shows that education and culture may be viewed as infrastructure, vital civic investments that have the ability to change the course of society. Second, multifunctionality is essential: in settings with limited resources, a single intervention must provide several benefits, including pride, safety, recreation, and education. Symbolism is important, third. Infrastructure that transports water or automobiles can be just as revolutionary as architecture that upholds dignity. However, these lessons need to be balanced with reality: famous enterprises can fail in the absence of consistent programming, inclusive governance, and dependable funding.

The takeaway is obvious: when limitations are reframed as design requirements, creative urbanization flourishes. Financial shortage, violence, and exclusion are not only reasons to do nothing; rather, they inspire new ideas for combining roles, fostering trust, and adding value. Medellín's Library Parks demonstrate how political bravery, and civic imagination can transform marginal hillsides into cultural hubs, even in towns that are stigmatized as hopeless. Although they are not flawless, they do demonstrate the capability.

As they wrap out this chapter and this section of the book, the Library Parks serve as a reminder that a city's future depends more on the scope of its vision than its financial resources. They serve as timeless lessons for both the Global North and the Global South: that resourcefulness is not a compromise but rather the cornerstone of urban development; that culture can be infrastructure; and that inclusivity can be planned.

References

1. Capillé C (2018) Political theatres in the urban periphery: Medellín and the library-parks project. Bitácora Urbano Territorial 28(2):125–134
2. Volkmann P (2018, October) The role of library parks as places of learning and culture in the transformation of the Learning City of Medellín, Colombia
3. Nevárez J (2020) Biblioteca España (Spain Library Park), Medellín, Colombia:"actions with my neighborhood". In: The urban library: Creative City branding in spaces for all. Springer International Publishing, Cham, pp 77–83
4. Davila V (2020) Lessons on how not to design a community space: the España Park public library in Medellín, a fading symbol of Hope. In: All-inclusive engagement in architecture. Routledge, pp 242–255
5. Capillé CC (2016) Spatial cultures of public libraries: architecture, collective use and political agendas in Medellín's library-parks. Doctoral dissertation, UCL (University College London)
6. Deluchi C (2020) The politics of social architecture in Medellín: a reading of the Parque Biblioteca España. Interstices:57–74
7. Garcia-Ramirez W (2023) Library parks in Colombia, or libraries in parks. Background of a political and architectural discourse. Arquitecturas del Sur 41(63)
8. Holmes JS, Amin Gutiérrez de Piñeres S (2014) Medellin's Biblioteca España: Progress in unlikely places. Available from https://utd-ir.tdl.org/items/097034e4-4d4f-41a6-bd44-a77ef9feedd6
9. Mangi E, Chen W, Heath T, Cheshmehzangi A (2024) Prioritizing sustainable urban regeneration practices: addressing contemporary urban challenges through the lens of public participation. In: Innovative public participation practices for sustainable urban regeneration. Springer Nature Singapore, Singapore, pp 1–11
10. Becker S (2013) The effects of the drug cartels on Medellín and the Colombian state. Doctoral dissertation
11. Cheshmehzangi A, You N, Allam Z, He BJ (2025) People-centered innovation: local actions for sustainable cities and communities. Springer Nature
12. Trupe L (2016) The effect of the Violence Prevention through Urban Upgrading (VPUU) intervention on violence-related injuries presenting to health facilities in Khayelitsha and Nyanga. Available from https://open.uct.ac.za/handle/11427/23709
13. Claghorn J, Orsini FM, Restrepo CAE, Werthmann C (2015) Rehabitar la Montaña: strategies and processes for sustainable communities in the mountainous periphery of Medellín. Urbe Rev Bras Gestão Urbana 8(1):42–60
14. Farieta MF (2023) Social Urbanism in Medellin–Integration of Landscape, Life, and Culture. Master's thesis, University of Maryland, College Park
15. Meninato P, Marinic G (2024) Medellín. In: Urban labyrinths. Routledge, New York, pp 49–72
16. Gandy M (2022) Natura Urbana: ecological constellations in urban space. MIT Press, Cambridge
17. Biczyńska E (2020) The Colombian city of Medellín: discussing the impact of social urbanism. Europa XXI 39:99–120
18. Fuentes Villabona MB (2017) The role of the shared public realm in social equality: Medellin and 'social urbanism'. Available from https://escholarship.mcgill.ca/concern/papers/9g54xh85d
19. Guzman-Rodriguez DE (2017) Dignity takings and dignity restoration: a case study of the Colombian land restitution program. Chi-Kent L Rev 92:871
20. Oyola SMR (2022) Uses of the concept of human dignity and the dignification of victims in transitional justice in Colombia. Eur Rev Int Stud 9(1):28–51
21. Zeldin DM (2011) Performing, vending, and walking in the city: the internally displaced population's occupation of spaces in Medellin, Colombia. Available from https://summit.sfu.ca/item/11692

22. Cheshmehzangi A (2023) City life experiences: lessons learnt for future urban regeneration strategies. In: Mapping urban regeneration: city life experiences in Yunnan, China. Springer Nature Singapore, Singapore, pp 239–246
23. Meninato P, Marinic G (2024) Transforming Medellín: architecture and urban design as agents of social change. J Archit Urban 48(1):83–90

Chapter 10
From Ingenuity to Transformation: The Future of Resourceful Urbanism

10.1 Translating Resourceful Urbanism Across Four Pillars

This book has suggested that resourceful urbanism is a systematic practice rather than a separate set of strategies. Its power is not found in discrete tactics, but rather in the ways that several strategies—such as circular resource management, adaptive reuse and retrofitting, community-led innovation, and multipurpose urban spaces—converge and support one another. The fundamental connection of these four pillars would be missed if they were treated as distinct categories. They weave flexibility, resilience, and regeneration into urban life like threads of a single fabric.

The autonomous foundation of resourceful urbanism is Community-Led Innovation [1]. The lived realities and expertise of city dwellers, especially those who have historically been excluded from planning procedures, serve as the foundation for change. However, community-led projects need infrastructures, methods, and venues that allow for local agency to be realized. The other pillars enter the picture at this point. Communities take the lead in bringing adaptive reuse and retrofitting to life by repurposing existing assets, such as abandoned rail lines, abandoned warehouses, or underutilized open spaces [2]. By transforming discarded or neglected assets into shared resources, these physical solutions express communal claims to the city and go beyond simple material efficiency.

Because reusing infrastructure and materials minimizes ecological loads, prolongs the life of resources, and closes waste loops, adaptive reuse and circular resource management naturally overlap. Large-scale industrial symbiosis in Kalundborg, Denmark, is an example of this synergy, while Pune's SWaCH cooperative shows how routine waste operations may be turned into sources of income, recognition, and environmental benefits. Without circularity, adaptive reuse runs the risk of becoming a one-time fix with no lasting impact. Every reuse is guaranteed to contribute to a larger metabolic shift, wherein labour, materials, and energy are conserved and recycled via urban systems, thanks to circularity.

A. Cheshmehzangi, *Resourceful Urbanism*, Urban Sustainability,
https://doi.org/10.1007/978-981-95-4138-6_10

Multifunctional urban spaces, the fourth pillar, connects the other pillars in terms of spatial form. Being resourceful means demanding that place provide several benefits at once, such as climatic resilience, cultural engagement, ecological restoration, and recreational activities. Multifunctionality guarantees that adaptive reuse promotes social and ecological enrichment in addition to structural survival. It makes it possible for community innovation to thrive in settings that support a range of users and uses. Additionally, it transforms circular flows into tangible, experiencing aspects of daily urban life, as shown in the network of Zoorasia and Forest Parks in Yokohama, where recreation, education, and biodiversity restoration all come together in one setting.

When combined, the four pillars break down silos within disciplines. They demonstrate that resilience encompasses cultural identity, civic trust, and adaptive governance in addition to ecological parameters and physical robustness. Regeneration results from the ongoing interaction of people, places, resources, and functions rather than from a single intervention [3]. Supportive infrastructures are necessary for a community cooperative to succeed; adaptive reuse only becomes transformative when it is integrated into larger resource cycles [4]; circular practices become popular when they are implemented in places that people value and occupy; and multifunctional landscapes endure when they are rooted in institutional support and community stewardship.

The way the four pillars contradict conventional development logics is equally significant. According to traditional paradigms, communities are portrayed as passive recipients of innovation, which is thought to be the result of capital-intensive technology. The appeal of new construction and demolition tends to overshadow adaptive reuse. Instead of being a social and political endeavour, resource management is presented as a technical efficiency. Instead of being a living experience, multifunctionality is reduced to master-planning platitudes [5]. These presumptions are reversed by resourceful urbanism. It maintains that communities are where innovation begins, that pre-existing assets are worth more than hypothetical blank slates, that circularity is both cultural and technical [6], and that multifunctionality is a design philosophy based on efficiency and happiness.

The temporal dimension of resilience [7] is also redefined by the combination of these four pillars. Resourceful urbanism encourages continuity and incrementalism rather than drastic one-time undertakings or temporary solutions. Through iteration, communities develop capacity; repurposed structures incorporate historical layers into future forms; circular flows create long-lasting feedback loops; and multipurpose spaces change to meet changing ecological and social demands. This is resilience as an active process rather than a list of defences.

Lastly, the concept of regeneration is expanded by these related tactics. Regeneration is the ability of cities to mend relationships—with their citizens, with their ecologies, and with their histories—rather than only ecological restoration or economic resurgence [8]. By valuing agency, community-led innovation mends societal rifts. Adaptive reuse revitalizes historic structures, repairing material legacies. By reestablishing the connection between production and consumption, circular management heals metabolic rifts. By giving various publics a common area, multifunctional spaces address spatial inequities. When combined, they create cities that not only endure within the bounds of the planet but also flourish by transforming limitation into opportunity.

The message becomes evident when interpreting resourceful urbanism across its four pillars: these are convergent lines rather than parallel tracks. Although each pillar is necessary, it is at their intersections that their full potential is shown. Adaptive reuse projects are given life by the democratic spirit of community-led initiatives; they are maintained by circular systems; and their reach into citizens' daily lives is expanded by multifunctional design. Therefore, resourceful urbanism is a model of integration rather than a toolbox of discrete strategies, where creativity emerges exactly from fusing disparate activities into a cohesive, regenerative whole.

10.2 Key Findings in Retrospect

The evidence is clear when examining the eight case studies and the four thematic pillars: resourceful urbanism is a living, developing practice rather than an ideal. Every case study has demonstrated how limitations, whether they be monetary, environmental, or spatial, may spur creativity rather than stand in the way of advancement. The crucial realizations that result can be grouped into four overarching lessons that, when combined, redefine the concepts of sustainability, resilience, and regeneration in urban areas.

10.2.1 Constraints as Engines of Innovation

The fact that scarcity is not a dead end is maybe the most compelling finding. Violence, neglect, and a lack of funding led to audacious expenditures in cultural and educational facilities in Medellín's Library Parks, which completely changed entire neighbourhoods. In Koroipita, cyclone vulnerability and land scarcity served as the foundation for the design of modular, cyclone-resilient housing that was constructed with sweat equity. These instances show how limitations compel towns and cities to adopt new perspectives, depart from traditional methods of growth, and recognize the worth of underutilized resources, abilities, and areas. Redefining constraint as a design brief allows for a level of dedication and creativity that is frequently stifled by excess.

10.2.2 Governance as Critical as Design

Another important realization is that governance models are just as important as actual interventions. The Coin Street Community Builders case demonstrated that the true innovation underlying the visible housing and public spaces is long-term stewardship organized through cooperative housing and a social enterprise model. In a similar vein, SWaCH in Pune has demonstrated that the survival of waste pickers and circular recycling processes is solely due to a hybrid governance model, which involves a cooperative that negotiates directly with homes and the municipality. These examples demonstrate how even the most impressive physical architecture can fail in the absence of strong, inclusive, and flexible governance. Resilience is maintained not just by structures and infrastructure but also by the organizations, agreements, and accountability systems that support them, according to resourceful urbanism.

10.2.3 Reuse and Circularity Beyond Technical Fixes

Adaptive reuse and circular resource management extend beyond the technical realm into cultural, social, and political domains. This is exemplified by the Goods Line in Sydney and Zeitz MOCAA in Cape Town: transforming abandoned rail tracks and grain silos into municipal and cultural treasures is about more than just embodied carbon conservation. It's about creating new kinds of belonging, claiming identity, and conserving memories. Similarly, Kalundborg's industrial symbiosis is not simply a technical exchange of steam, water, and gypsum—it is a trust-based network built over decades that illustrates how circular flows require governance, negotiation, and shared values as much as pipelines and treatment plants. Rather than being limited to efficiency measures, circularity thrives when it is ingrained in social contracts and civic imagination.

10.2.4 Multifunctionality Unlocks Efficiency and Delight

The fourth realization is that value is multiplied by multifunctionality. Multipurpose urban areas demonstrate how one intervention can address several objectives at once, including climate adaptation, cultural programming, public recreation, and ecological restoration. The Zoorasia and Forest Park system in Yokohama serves as an example of how peri-urban space may simultaneously provide recreational possibilities, biodiversity advantages, and climatic co-benefits. Similar to this, Medellín's Library Parks serve as municipal plazas, schools, cultural hubs, and dignified emblems all wrapped into one infrastructure. Multifunctionality is a concept of efficiency and enjoyment rather than just a design approach, making sure that limited resources and available space are used to their fullest potential.

These results collectively suggest a significant rethinking of urbanism. They demonstrate that resourcefulness is about invention, inclusivity, and regeneration rather than austerity or compromise. Multifunctionality enables cities to flourish within constraints while enhancing civic life; circularity is as cultural as it is technical; governance may be as transformative as architecture; and constraints can become catalysts for innovation.

These observations serve as a reminder that the people who can best mobilize relationships, histories, and modest resources to create long-lasting public value will shape urbanism's future, not those who possess the greatest money or technology. Resourceful urbanism is an advanced paradigm for all communities facing planetary restrictions, not a backup plan for the under-resourced.

10.3 Global Lessons from Eight Cases: Convergences, Divergences, and Shared Pathways

The eight case studies in this book were specifically selected to cover both the Global North and the Global South. They include London's Coin Street, Sydney's Goods Line, Medellín's Library Parks, Pune's SWaCH cooperative, Kalundborg's industrial symbiosis, Yokohama's multifunctional landscapes, Cape Town's Zeitz MOCAA, and Fiji's Koroipita model towns. They are located in wildly disparate political economies, cultural settings, and geographic locations. Nevertheless, when combined, they provide a logical account of the new paradigm of resourceful urbanism. They show that although situations influence methods, the ideas of creativity, circular invention, adaptive reuse, and community leadership are universal.

10.3.1 Constraints as a Universal Catalyst

The first and most obvious lesson is that the Global South is not the only region with limits. The case studies show how the Global North faces its own kind of constraints, such as abandoned infrastructures, environmental limitations, and social fragmentation, even if the lack of capital, land, and governmental capacity is frequently linked to developing situations. The Goods Line in Sydney was created not because of unrestricted funding but rather because it was realized that repurposing an abandoned rail track was more efficient and less expensive than destruction and redevelopment. Coin Street in London was successful not because developers made quick money but rather because locals banded together to fight speculative influences and establish common land ownership. Similar to this, communities in Fiji and Pune were compelled to innovate through sweat equity, decentralized waste systems, and incremental infrastructure due to financial limitations, disaster susceptibility, and state indifference. Constraint functions as a catalyst for imaginative reimagining rather than as a barrier in all instances.

10.3.2 Governance as the Decisive Arena

A second lesson is that governance systems, whether they be industrial networks, cooperatives, or municipal-community partnerships, can have more sway than design itself. The formalization of waste pickers in Pune through SWaCH only persisted because accountability procedures were incorporated into governance agreements that acknowledged their role. Because businesses, towns, and farmers institutionalized trust and collaboration through contracts, industrial symbiosis flourished in Kalundborg. The Library Parks in Medellín were significant not just as architectural landmarks but also as tools of a new municipal ideology that placed an emphasis on education and culture as infrastructure. In contrast, Zeitz MOCAA in Cape Town raises concerns about accessibility and representation and highlights the dangers of governance models that rely too heavily on private endowments and philanthropy. The common lesson is that resourceful urbanism is as much about who governs and how, as it is about what is built.

10.3.3 Reuse, Circularity, and the Value of the Existing

The amazing possibility of reusing resources, infrastructure, or social processes is another lesson we have in common. Instead of being destroyed, Cape Town's grain silo was transformed into Africa's largest museum of contemporary art, preserving embodied energy while changing identity. In Sydney, integrating rail tracks into the pavement of the Goods Line became a symbolic act of industrial memory continuity. Industrial waste was recycled into resources in circular flows in Kalundborg, yielding both financial and environmental benefits. In Pune, on the other hand, the circularity of waste pickers' labor changed the material metabolism of the city while fostering livelihoods and dignity. Together, these instances refute the conventional wisdom that progress necessitates beginning over; rather, they show that it is more cost-effective and environmentally friendly to build with what already exists, both socially and physically.

10.3.4 Multifunctionality as Efficiency and Delight

The power of multifunctionality is an additional lesson. When systems, infrastructures, and spaces fulfill several purposes at once, resourceful urbanism flourishes. Medellín's Library Parks served as plazas, schools, cultural centers, and dignified emblems in addition to libraries. The Zoorasia and Forest Park network in Yokohama is more than just a zoo; it's also a recreation area, a climate regulator, an outdoor classroom, and a regenerated satoyama landscape. In addition to housing, Koroipita is a community that values education, recycling, and resilience. Multifunctionality creates areas of happiness and a sense of community while maximizing the public value of scarce resources.

10.3.5 Convergences Across Contexts

There are notable similarities between the Global North and South examples when compared side by side. First, every case exemplifies a refusal to squander resources, labour, or land. Resourceful urbanism thrives on preserving and increasing value, whether it is in Pune's recycling cooperatives, Denmark's industrial symbiosis, or London's communal housing. Second, every example demonstrates the value of incrementalism: whether it is Coin Street's periodic land acquisitions or Koroipita's staged housing expansion, change frequently happens gradually. Third, the cases show that resourcefulness is inextricably linked to culture, whether it be in the form of art, identity, or dignity. Regeneration is always cultural rather than merely technological.

10.3.6 Divergences Shaped by Context

However, important divergences exist. As demonstrated in Pune, Medellín, and Lautoka, resourceful urbanism frequently emerges from grassroots necessity in the Global South, where communities drive innovation in the lack of governmental provision. With design studios, philanthropic foundations, or municipal utilities playing key roles, innovative ideas, on the other hand, frequently arise within or alongside stronger state and institutional frameworks in the Global North. Additionally, there are differences in financing models: instances from the Global North typically rely on governmental budgets, endowments, or corporate partnerships, whereas those from the Global South mostly rely on sweat equity, donor support, or cooperative arrangements. However, these differences are more about initial circumstances than they are about talent; in both cases, resourcefulness steps in when traditional methods fall short.

10.3.7 Universal Insights for a World in Crisis

The main takeaway is that resourceful urbanism is relevant since today's crises—such as resource depletion, inequality, and climate breakdown—cut beyond national boundaries. No city, North or South, can afford to ignore neighbourhoods or waste resources. The eight examples illustrate universal truths: multifunctionality enables cities to thrive within limitations; circularity is cultural as much as technical; existing assets retain unrealized potential; and communities are essential co-creators.

10.3.8 Beacons for Transformation

Lastly, these events ought to be interpreted as beacons rather than as isolated experiments. They illuminate paths that other cities, confronting their own risks and shortages, can take. In speculative markets, Coin Street demonstrates the potential for collective land stewardship. Concrete can be as infrastructural as culture, as Medellín shows. Kalundborg demonstrates that industries may work together for the good of everybody. Pune serves as an example of how informality may be formalized into sustainability and dignity. The Goods Line in Sydney is a classic example of low-cost, high-impact repurposing. Koroipita shows how disaster-prone areas can gradually increase resilience. The multipurpose landscapes of Yokohama serve as an example of how peri-urban land may benefit both the people and the climate. Despite its criticisms, Cape Town's Zeitz MOCAA serves as an example of the symbolic value of transforming industrial artifacts into cultural futures.

Collectively, they demonstrate that resourceful urbanism is a paradigm that spreads by principles rather than recommendations rather than a single recipe. It is more about understanding how creativity under pressure can provide long-lasting societal benefit than it is about reproducing exact models. These examples make a strong case for cities all around the world that are juggling danger, scarcity, and change: the future will be created by ingeniously repurposing what already exists rather than starting from scratch.

10.4 A Paradigm Shift in Motion

When we talk about resourceful urbanism, we are talking about a fundamental change in the way we think about, plan, and manage cities [9, 10]. Planetary boundaries are causing the traditional models that have dominated urban thought and practice for decades—techno-utopianism on the one hand, and growth-centric development on the other—to falter. They make the assumption that there is plenty of money, energy, and political will. They see cities as either everlasting growth engines or as testing grounds for digital optimization. However, neither framing has produced the resilient, inclusive, and regenerative futures that we so desperately need. In this context, resourceful urbanism shows up as a compelling alternative paradigm that is already in motion, rather than as a fringe activity.

The "smart city" story exemplifies techno-utopianism, which promises salvation via sensors, algorithms, and seamless communication [11]. The city as an operating system, with behaviours optimized and inefficiencies debugged, has a compelling charm. However, there is a deep vulnerability hidden beneath the hyperbole. Visions for smart cities are capital-intensive, frequently exclusive, and more likely to reinforce inequality than to eliminate it. They put efficiency measures ahead of ecological restoration, proprietary platforms ahead of civic agency, and data streams ahead of interpersonal relationships. Technologies' promises crumble and reveal a hollow core when they fail, which is inevitable.

In contrast, growth-centric urbanism maintains that expansion is necessary for prosperity [12]. Mega-projects, speculative redevelopment, and cycles of demolition and reconstruction are its mainstays. This reasoning is just as fragile. It uproots vulnerable communities, consumes resources and land at unsustainable rates, and maintains linear waste and extraction processes. Constraints are framed by growth-centric paradigms as challenges to be overcome rather than as realities to be confronted [13]. They waste the very resources that may serve as the cornerstone of regeneration in the process—community ties, cultural memory, and ecological systems.

By redefining communities as co-creators and restrictions as design briefs, resourceful urbanism challenges both of these ideas. Resourceful urbanism prioritizes what already exists, such as underutilized spaces, forgotten infrastructure, local skills, and cultural memory, while techno-utopianism looks to imported technologies. Resourceful urbanism views scarcity as a chance to innovate, reuse, and multiply value through multifunctionality, whereas growth-centric models view it as a problem. Reusing a rail corridor, formalizing waste pickers' knowledge, or turning industrial byproducts into shared resources are all examples of adaptation. It is not a backup plan; rather, it is the main force behind change.

This change is both pragmatic and intellectual. According to resourceful urbanism, cities are living ecologies where social, cultural, and material systems coexist rather than being machines to be optimized or commodities to be exchanged [14]. It recognizes that resilience is developed via routine maintenance, repair, and care procedures rather than just in labs or boardrooms. It acknowledges that the ability to rethink relationships—with land, with materials, and with each other—is where innovation starts, not an abundance of resources.

Importantly, creative urbanism emphasizes communities as co-creators [15]. This is local agency-based governance and design, not participation as token consultation. Medellín's Library Parks teach us that when culture is in line with community needs, it becomes infrastructure, not just that libraries lessen violence. The respect that waste pickers receive when they are acknowledged as participants in the city's metabolism is just as important to Pune's SWaCH cooperative's success as the tons of waste that are diverted. Projects like London's Coin Street show how locals may claim and manage property against speculative pressures, creating futures that are more in line with collective than individual gain, even in the Global North, where institutions are more robust.

Additionally, the paradigm shift reinterprets what constitutes innovation [16]. Resourceful urbanism acknowledges inventiveness in sweat work, incrementalism, and the layering of uses and meanings rather than equating innovation with the newest technology. Whether rehabilitating Yokohama's satoyama landscapes or bringing pedestrian life to Sydney's Goods Line, it finds value in actions that may seem simple but result in long-lasting, regenerative effects. These are the subtle revolutions that keep cities afloat for decades, not the dazzling displays of urban branding.

This paradigm shift is in motion and presents problems to citizens, legislators, and experts alike. For professionals, it demands humility: to design with what exists, to learn from informal systems, to embrace adaptation over replacement. It takes bravery on the part of legislators to support community-driven projects, enact laws that encourage reuse rather than demolition, and gauge success in resilience and wellbeing rather than just GDP growth. It encourages residents to exercise agency by viewing themselves as contributors to the development of their cities rather than as recipients of municipal services.

This change is already happening, as demonstrated by the eight case studies in this book. They demonstrate that resourceful urbanism is a real practice across continents rather than a theoretical ideal. Integration between the formal and informal, the ecological and the human, and the local and the global is its guiding principle. It offers a regenerative city, one that can fix the past, adjust to the present, and support the future, rather than a flawless one.

10.4.1 Five Pathways to Designing Regenerative Cities Through Adaptive Reuse and Circular Innovation

If resourceful urbanism is a paradigm shift, then putting ideas into action is what gives it life. Cities must operationalize previous innovations into guiding principles for everyday living, government, and design, rather than merely applauding their inventiveness. Out of the lessons learned in this book, five approaches or pathways stand out: flexible tactics based on the reality of scarcity, creativity, and teamwork, rather than strict prescriptions (see Fig. 10.1).

1. **Reclaiming Governance**

   The first pathway demands a shift in how cities are governed. When governance is hybrid, participatory, and based on community stewardship, *resourceful urbanism* flourishes. This entails bolstering systems that put locals at the center of decision-making rather than on the outskirts. While Pune's SWaCH cooperative shows how waste pickers, who were once stigmatized, may become formal actors in urban systems, London's Coin Street exemplifies the resilience of community trusts.

   New tools are needed to reclaim governance: cooperatives to manage services, community land trusts to protect affordability, participatory councils to resolve disputes, and hybrid partnerships that unite private players, governments, and civil society. Without this component, resourcefulness runs the risk of being appropriated into hierarchical tactics that perpetuate exclusion. Democratic resilience, or governance that adjusts, incorporates, and safeguards community agency against the volatility of markets and politics, is the cornerstone of this road.
2. **Building with What Exists**

   The second pathway is material and spatial, emphasizing creative reinvention, adaptive reuse, and retrofitting above demolition and unrelenting growth. Energy, labour, and memory are all embodied in every building, hallway, and piece of infrastructure. Destroying and rebuilding is frequently expensive, waste-

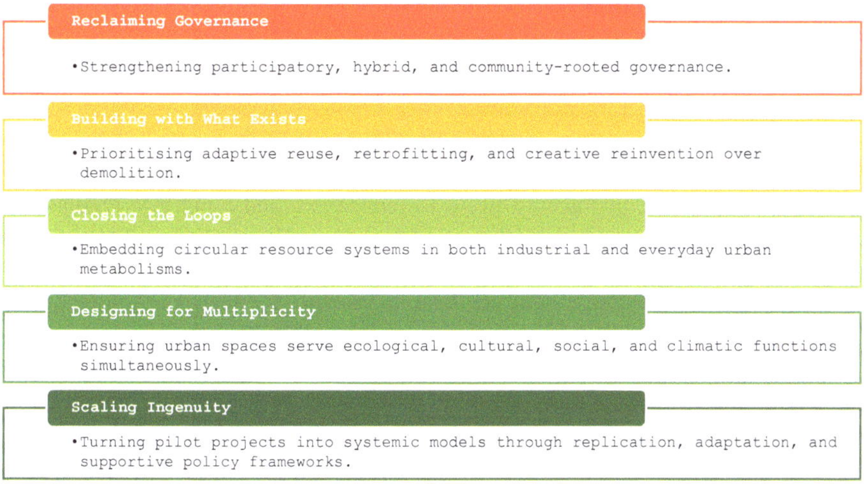

Fig. 10.1 Summary of five pathways to designing regenerative cities through adaptive reuse and circular innovation

ful, and disruptive to society. Both Cape Town's Zeitz MOCAA and Sydney's Goods Line serve as examples of how abandoned industrial artifacts can be restored to serve as municipal and cultural assets without destroying their past.

According to this approach, cities must be planned like archaeological strata, with each intervention recognizing the past while paving the way for the future. Reusing unused lots for community projects, retrofitting buildings for climate efficiency, and adding new uses to existing buildings all assert continuity while conserving resources. According to this vision, "newness" comes from the art of reinvention rather than from a complete overhaul.

3. **Closing the Loops**

   Circular resource flows are essential to resourceful urbanism; what one system discards, another must appreciate. The template is the Kalundborg industrial symbiosis, in which waste products, gypsum, water, and steam are shared between businesses and families, lowering expenses and emissions. Circularity, however, is not limited to industry; it also applies to normal metabolisms. Pune's decentralized networks for recycling and composting show how homeowners and unpaid laborers take on the role of stewards of loops that keep garbage out of landfills and lower carbon footprints.

   Closing loops involves social contracts—agreements that specify accountability, responsibility, and trust—as well as infrastructure, such as pipelines, treatment plants, and composting facilities. By integrating cycles of repair, recovery, and reuse into the very fabric of urban systems, it dispels the idea of disposability. In the end, circularity reframes cities as metabolic creatures with the ability to regenerate themselves rather than as linear engines of consumption.

4. **Designing for Multiplicity**
   Making sure that each intervention fulfills several functions, including ecological, cultural, social, and climatic, is the fourth pathway, which is spatial. Multifunctionality is essential in a society where resources are becoming scarcer and urban populations are growing. The Zoorasia and Forest Park in Yokohama serves as an example of how peri-urban area may serve as a habitat for biodiversity, a place for enjoyment, a place for education, and a place to regulate the climate. In congested hillside barrios, Medellín's Library Parks exemplify this same idea: libraries serve as municipal anchors, plazas, and classrooms.
   Multiplicity design is both practical and poetic. It enhances experience and maximizes efficiency. It creates areas—often within the same square meter—where kids play, seniors congregate, ecosystems recover, and communities learn. The idea is straightforward: instead of dividing up purposes, cities should create spaces that recognize the interdependence of ecological and human life.
5. **Scaling Ingenuity**
   Trajectory is the subject of the last pathway. Innovative endeavours frequently start off as pilots, small-scale tests that demonstrate what is feasible. But their ability to become systemic models is what gives them strength. Originally a small collection of cyclone-resistant houses, Koroipita, Fiji, is today known throughout the Pacific as a model village. The waste-pickers' cooperative in Pune began as a union fight before developing into a municipal service that now serves millions of people. Scaling ingenuity is not about replication without context but about adaptation—translating principles into diverse conditions.
   Cities require enabling policies, such as training initiatives, social impact bonds, seed grants, and supportive regulatory frameworks, in order to do this. Additionally, they require learning networks that link initiatives across national boundaries so that practices in Nairobi or Jakarta can benefit from lessons learned in Copenhagen or Medellín. In order for resourceful urbanism to transition from exceptions to norms, scaling serves as the pivot between innovation and transformation.

10.4.2 From Pathways to Paradigm: Linking Back to the Advanced City

When combined, these five routes provide a framework for regenerative urban areas. They show how space, resources, culture, and governance must all be mobilized simultaneously to create systemic resilience rather than transient resilience. Each pathway is a call to consider government as inclusion, to view buildings as resources, to approach spaces as multifunctional, to design metabolisms as circular, and to scale innovation as systemic transformation.

These pathways also intersect with the idea of the "Advanced City", a conceptual framework I have developed that reimagines urbanism as a choice to evolve beyond incremental sustainability [17]. The Advanced City is resourceful, regenerative, and relational; it is not only clever or green. These five pathways—community-based advanced governance; adaptive reuse for advanced material use; circular flows for advanced metabolisms [18]; multiplicity-designed spaces; and advanced transitions that scale innovation systemically—all directly align with this vision.

The convergence is instructive. The Advanced City concept and resourceful urbanism are two sides of the same revolution, not two distinct schools of thought. Both maintain that developing creativity, resiliency, and relational intelligence—rather than acquiring more cash or technology—is what cities ought to focus on in the future. Both contend that limitations should inspire creativity rather than serve as justifications for paralysis. Both call for cities to transform into regenerative systems that can fix the past, adjust to the present, and maintain the future.

In this way, the five pathways represent stepping stones toward the Advanced City itself rather than merely practical suggestions. They encourage communities, legislators, and planners to view resourcefulness as a philosophy and a strategy, to recognize design potential in every limitation, and to acknowledge that the most developed cities will be those that prosper not because of scarcity but because of it.

10.5 Innovation and Integration as the Way Forward

The word *innovation* has been hijacked by the spectacle of technology. All too frequently, contemporary urbanism is associated with shiny smart districts, autonomous vehicles/cars, and computerized dashboards that are more for branding than for actual resilience. However, the examples analyzed in this book demonstrate that the most revolutionary inventions are rarely striking. They are peaceful, relational, and integrative. The ability to integrate governance, materials, ecosystems, and daily routines into coherent, regenerative systems is what makes resourceful urbanism innovative, not the new gadgets [19]. The most radical innovation of all is integration itself.

Consider Kalundborg's industrial symbiosis. Its genius lies in integrating businesses, farms, and municipal utilities into a common metabolism rather than in futuristic hardware. Steam, water, gypsum, and sludge loops are closed, turning waste into resource. Although the technical infrastructure is important, the social and institutional integration that upholds trust over many years is the true innovation. Similar to this, the SWaCH cooperative in Pune shows how families, waste pickers, and local government agencies may all be integrated into a circular service system. The true breakthrough is the act of integrating formal and informal, labor and recognition.

Integration also defines the spatial dimensions of resourceful urbanism. The layered design of Medellín's Library Parks—architecture combined with education, public plazas combined with civic dignity, and culture incorporated into security measures—is what makes them noteworthy rather than their technical innovation. In terms of ecology, the Zoorasia and Forest Park network in Yokohama exemplifies the same idea by combining community management, recreation, flood control, and biodiversity restoration. These examples demonstrate that the total of the pieces is far less than the effect of connecting systems, actors, and scales.

What implications does this have for cities' future? It implies that the quest for integrative capacity rather than high-tech spectacle will be the next frontier of urbanism. Cities need to learn how to integrate technology with community agency, link material reuse with cultural memory, and match governance processes with ecological cycles. In this context, innovation is more about orchestration than creativity. It involves creating interfaces that permit the productive circulation of energy, matter, and meaning between the public and private spheres, as well as between formal and informal and human and non-human.

Humility is also a prerequisite for integration. It acknowledges that no one actor—not the state, a business, or a community—has all the answers. Rather, resilience arises from the coordination, negotiation, and co-creation of various players. This calls for organizations to share authority, planners to learn from informality, and designers to view limitations as opportunities for creativity. In and of itself, integration governance is a challenging innovation that calls into question long-standing privileges and hierarchies.

Lastly, scale is redefined through integration. It is impossible to comprehend a regenerative city solely at the level of its infrastructure or buildings. It must be seen as a multi-scalar network in which peri-urban forest parks anchor climate resilience for entire regions, where neighbourhood composting links to metropolitan food systems, and where renovated warehouses link to regional cultural economies. Innovation is found in the connective tissue that holds each component together rather than in each component alone.

The way forward, then, is not to chase the illusion of technological salvation but to cultivate integration as practice and ethos [20]. The most successful cities in the twenty-first century will not be the most technologically sophisticated or the most brilliantly reconstructed; rather, they will be those that successfully integrate people with place, systems with systems, and the past with the future. Integration is the real architecture of regeneration, and innovation is integration.

10.6 Concluding Reflections: A Brief and Bold Closing

This book's journey through eight cases and four thematic pillars has led to the straightforward but profound conclusion that resourceful urbanism is the seed of a new paradigm rather than an afterthought or a marginalized group's survival strategy. In environments as disparate as London and Lautoka, Pune and Kalundborg, Medellín and Yokohama, we have witnessed how creativity flourishes in the furnace of limitation rather than plenty. When resources are restricted, creativity flourishes; when capital is constrained, solidarity grows; and when risk is high, resilience is created.

Therefore, resourceful urbanism is not a band-aid solution. A regenerative revolution that reinterprets the very fabric of cities is about to begin. It invites us to judge progress by the longevity of public value provided, the inclusivity of governance structures, the inventiveness of reuse, and the ability to regenerate natural and social systems rather than the magnitude of capital deployed or the spectacle of new technologies. It maintains that thriving within constraints is not a compromise but rather the hallmark of sustainability, and that accomplishing more with less is not austerity but ingenuity.

The four pillars explored in this book, i.e., *Community-Led Innovation, Adaptive Reuse and Retrofitting, Circular Resource Management, and Multifunctional Urban Spaces*, are not isolated tactics but rather work in concert with one another. When taken as a whole, they reinterpret resilience as a process of relation and regeneration. They demonstrate that multifunctionality increases value across ecological, social, and spatial dimensions, that governance is just as important as design, and that circularity is as cultural as it is technical. These are lived practices that have been demonstrated in a variety of political economies and geographical locations, not just theoretical concepts.

The eight case studies have shown how effective this strategy is. From Koroipita's sweat equity to Medellín's Library Parks' monumental dignity, from Kalundborg's industrial symbiosis to Pune's waste cooperatives, from Sydney's Goods Line to Cape Town's silo-turned-museum, from London's Coin Street community stewardship to Yokohama's multipurpose landscapes, each project demonstrates that resourceful urbanism is not only feasible but is already taking shape. They serve as examples of what cities may become when they approach communities as co-creators, embrace constraint as a design brief, and regard the underutilized and underappreciated as the cornerstones of revitalization.

But this book is not just analysis; it is also invitation. It instructs the planner to consider the current framework when developing and to look beyond the master plan [21]. It suggests that legislators encourage cooperatives, pass legislation that promotes reuse, and measure success based on well-being and resilience rather than merely growth. The architect should consider memory and material as resources and plan for multiplicity rather than singularity. In contrast to being a consumer, it tells the citizen to establish their role as a steward, innovator, and citizen of their city.

The stakes are global, so the invitation is urgent. Resource depletion, social injustice, and climate change are not hypothetical dangers [22]; rather, they are experienced realities that influence every urban vista. The outdated ideologies—growth-centric urbanism, which demands unending expansion, and techno-utopianism, which promises deliverance through technology—are worn out. Neither can point us in the direction of just, resilient, and environmentally conscious futures. A third route—a paradigm change that is both practical and imaginative—is provided by resourceful urbanism. Because it makes use of what already exists and draws on the institutional and practical realities of cities, it is pragmatic. It is visionary because it sees regeneration as the potential for happiness, dignity, and solidarity within the bounds of the earth, rather than just survival.

In the future, the question is not whether resourceful urbanism can be successful—it has innumerable examples—but rather if we will decide to make it the rule rather than the exception. This decision will be influenced by community-empowering governance structures, circularity and reuse-promoting policies, designs that embrace diversity, and a shared desire to expand creativity across boundaries and situations.

Therefore, this book ends with momentum rather than finality. Every reader is a part of the revolution that resourceful urbanism is already bringing about. The challenge is to cultivate it, expand it, and include it into the routine tasks of creating, managing, and residing in cities. It is an exhortation to view resourcefulness as ability rather than compromise, as the art of regeneration rather than the management of scarcity.

Those who trust that technology will save us alone, or who want to delete and rebuild, will not write the future of cities. It will be written by people who recognize the potential for rejuvenation in every underutilized material or space, the possibility of design in every limitation, and the seed of governance in every community. This journey's compass is resourceful urbanism. How bravely we will accept it is the question, not if it will matter.

References

1. Crowley D, Marat-Mendes T, Falanga R, Henfrey T, Penha-Lopes G (2021) Towards a necessary regenerative urban planning. Insights from community-led initiatives for ecocity transformation. Cidades. Comunidades e Territórios, (Sp21)
2. Cheshmehzangi A (2025) Vision and strategy in advanced cities: catalyzing growth, participation, innovation, and communication. In: Advanced City: for the future of cities and opposing smart cities. Springer Nature Singapore, Singapore, pp 281–291
3. Moya ARZ, Yáñez CJN (2017) Impact of area regeneration policies: performing integral interventions, changing opportunity structures and reducing health inequalities. J Epidemiol Community Health 71(3):239–247
4. Pileggi JR (2024) Dilapidated to vibrant: adaptive reuse as a catalyst for regenerating urban areas through public private partnerships. Doctoral dissertation, Kansas State University
5. Ghafouri A (2016) Sustainable urban form: multifunctionality and adaptation: redefining urban spaces as multifunctional shared areas. Doctoral dissertation, Université de Strasbourg
6. Reyneke P (2024) An ethnographic study of outside-circularity and deconstructive creation from the waste reuse practices of the urban waste precariat. Doctoral dissertation, University of the Witwatersrand, Johannesburg
7. Erik Johansson BJ, Lundberg J (2017) Resilience and the temporal dimension–the chimera of timely response. Theor Issues Ergon Sci 18(2):110–127
8. Fusco Girard L, Nijkamp P, Nocca F (2023) Reconnecting the city with nature and history: towards circular regeneration strategies. Available from www.torrossa.com/gs/resourceProxy?an=5650226&publisher=FM0520
9. Agudelo-Vera CM, Mels AR, Keesman KJ, Rijnaarts HH (2011) Resource management as a key factor for sustainable urban planning. J Environ Manage 92(10):2295–2303
10. Rae DW (2008) City: urbanism and its end. Yale University Press, London
11. Cheshmehzangi A (2025) Advancing sustainably, nurturing intelligently: calling for the development of 'brother cities' and 'advanced development goals (ADGs)'. In: Advanced City: for the future of cities and opposing smart cities. Springer Nature Singapore, Singapore, pp 481–494

12. Savini F, Ferreira A, von Schönfeld KC (2022) Uncoupling planning and economic growth: towards post-growth urban principles: an introduction. In: Post-growth planning. Routledge, pp 3–18
13. Lennox E, Hollender R (2020) Alternatives to growth-centric development. An ECI teaching module on social and economic issues. 53 p, published by economics in context initiative at Boston University, Boston, MA. Available from https://www.bu.edu/eci/files/2020/01/Alternatives-to-Growth_final.pdf
14. Swyngedouw E (2006) Circulations and metabolisms:(hybrid) natures and (cyborg) cities. Sci Cult 15(2):105–121
15. Mengi O (2020) Creative placemaking revisited: exploring major drivers for the practice of making and co-creation. Int J Knowl-Based Dev 11(3):220–243
16. Cheshmehzangi A, You N, Siri J, Birch E (2024) The nexus between urban innovation and sustainable development goals (SDGs). In: Harnessing urban innovation to unlock the sustainable development goals. Springer Nature Singapore, Singapore, pp 1–15
17. Cheshmehzangi A (2025) A new thinking: the idea of Advanced City thinking. In: Advanced City: for the future of cities and opposing smart cities. Springer Nature Singapore, Singapore, pp 1–9
18. Papageorgiou A, Björklund A, Sinha R (2024) Applying material and energy flow analysis to assess urban metabolism in the context of the circular economy. J Ind Ecol 28(4):885–900
19. Girard LF (2021) The evolutionary circular and human centered city: towards an ecological and humanistic "re-generation" of the current city governance. Hum Syst Manag 40(6):753–775
20. Munson TJ (2002) Technologies of sin & salvation: the search FOR contemporary cognitive-aesthetic technologies in the age of the perpetual innovation economy. Doctoral dissertation, Georgetown University
21. Cheshmehzangi A (2025) Urban planning for the contemporary age: navigating complexities and shaping urban futures. Encyclopedia 5(1):19
22. Blomfield M (2015) Climate change and the moral significance of historical injustice in natural resource governance. The ethics of climate governance 3–22

MIX
Papier aus verantwortungsvollen Quellen
Paper from responsible sources
FSC® C105338

If you have any concerns about our products,
you can contact us on
ProductSafety@springernature.com

In case Publisher is established outside the EU,
the EU authorized representative is:
Springer Nature Customer Service Center GmbH
Europaplatz 3, 69115 Heidelberg, Germany

Printed by Libri Plureos GmbH
in Hamburg, Germany